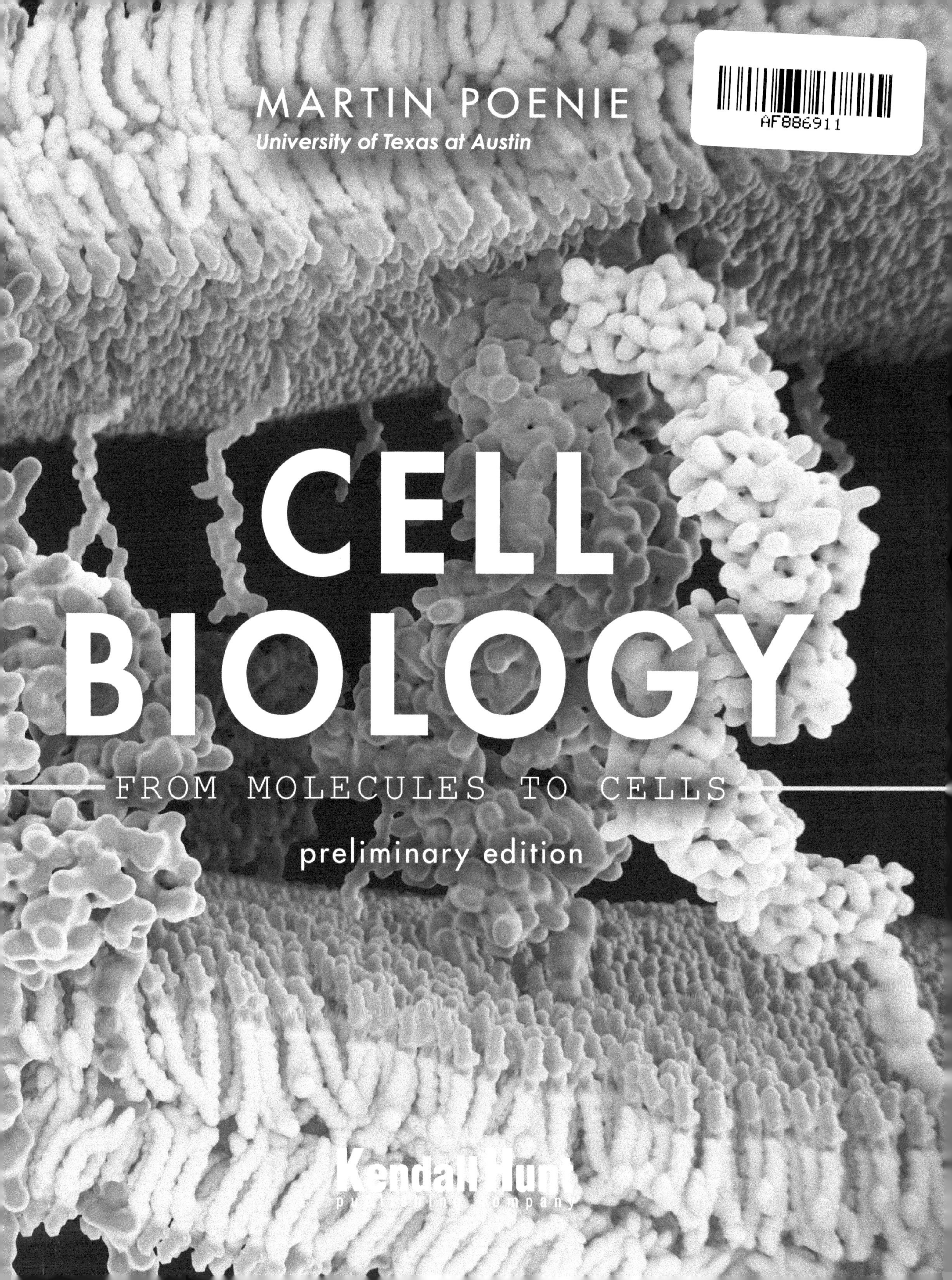

MARTIN POENIE
University of Texas at Austin

CELL BIOLOGY

FROM MOLECULES TO CELLS

preliminary edition

Kendall Hunt
publishing company

Cover image: © Shutterstock.com

Kendall Hunt
publishing company

www.kendallhunt.com
Send all inquiries to:
4050 Westmark Drive
Dubuque, IA 52004-1840

ISBN 978-1-5249-8163-1

Published in the United States of America

Contents

Beginnings

1

1 What Do We Know about the First Living Things?

The origin of life from a scientific perspective is one of the most mysterious and difficult problems in science. It is a question plagued by almost insurmountable problems. One of the problems is, given that the oldest life was microbial, these organisms leave very little by way of fossils. One can resort to biochemical markers but these are fraught with interpretive problems. Second, the oldest rocks on earth are about four billion years old and these are very rare. Thus much of earth history as coded in the rocks has been erased through events that continually renew the earth's surface. So, while we cannot provide a definitive answer, we can survey what we do know.

The earth formed about 4.5 billion years ago at the same time the sun was forming. It was originally thought that for the first half billion years, the earth was a fiery inferno, so this period was named the Hadean. Recent evidence from tiny crystals of zirconium silicate, known as zircons, has changed this picture. When zircons crystalize, they can incorporate uranium into the crystal lattice but not lead. This makes them ideal geochronometers because any lead in the crystal can be assumed to come from the slow radioactive decay of uranium trapped in the crystal. Silicate (SiO_4) in the zircon contains four oxygen atoms, and geologists can tell from the oxygen isotope composition (^{16}O or ^{18}O) whether the zircon crystal formed at extremely high temperatures deep in the earth or in cooler wet environments. Scientific studies have shown that zircons dated 4.4 billion years ago indicate that there was water on earth that far back (Mojzsis, Harrison, and Pidgeon 2001; Valley, Peck, and King 2002; Wilde et al. 2001). Earth may have been hot due to collisions with other planetoids such as that hypothesized to have formed the moon, volcanic activity, and bombardment with huge asteroids, so the water might have been steam, but in one form or another it would have been present.

A. Timeline of Events Related to Life on Earth

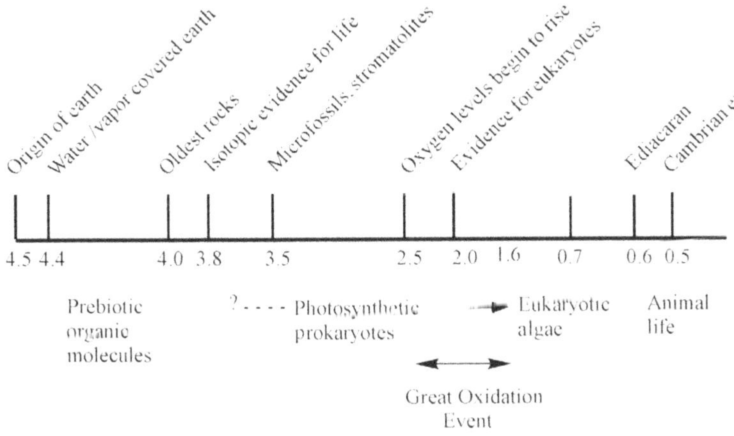

Figure 1.1. Timeline of events related to life on earth.

Source: M. Peonie.

The first evidence for life on earth comes from studies of carbon in rocks that date from 3.7 to 3.8 billion years ago. The carbon was subject to extreme heat and pressure causing it to liquefy as graphite before cooling. Thus there is no telling what its original form was or what living things gave rise to it. However, there are signatures in the stable carbon isotopes present in the rocks indicating that originally, the carbon was derived from living organisms. Carbon in the atmosphere can be found in three isotopic forms: ^{12}C, ^{13}C, and ^{14}C. Of these, ^{14}C is radioactive and decays relatively quickly, so it is not relevant to these studies. Both ^{12}C and ^{13}C are stable. A number of different organisms, including plants and photosynthetic bacteria (cyanobacteria), can take in carbon dioxide from the atmosphere and turn it into molecules the cells can use. However, these organisms greatly prefer carbon dioxide made from ^{12}C as opposed to ^{13}C. The ^{12}C to ^{13}C ratio in carbon deposits can therefore serve as a marker for carbon that is derived from living organisms.

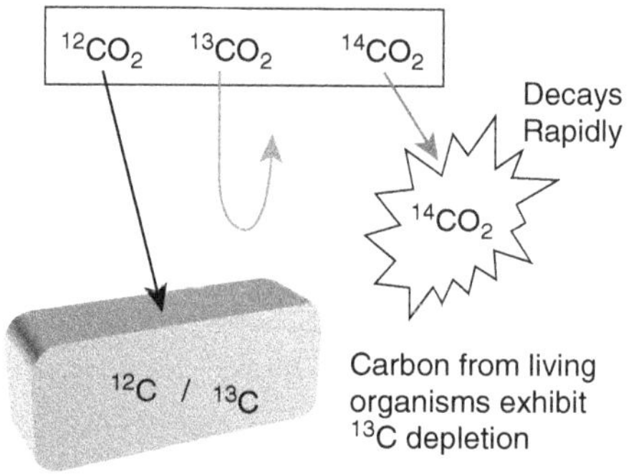

Figure 1.2. Carbon isotopes.

B. Isotopic Evidence for Life at 3.8 Billion Years?

Using this principle, scientific studies have shown that carbon deposits in rocks that are 3.7–3.8 billion years old show ^{12}C to ^{13}C ratios that indicate they are derived from living things. This constitutes the earliest evidence for live on earth.

2 What Were the First Cells Like?

While we do not know what organisms generated the graphite globules that provide evidence of the first life on earth, another recent study might give us a clue. We might begin to understand what the original cells were like by looking at stromatolites from Greenland (Nutman et al. 2016). Stromatolites are rocklike mounds formed by microbial communities and in particular, cyanobacteria. Cyanobacteria (blue-green algae) are especially interesting because they are photosynthetic. This recent but controversial study demonstrated that there are stromatolites in Greenland that are 3.7 billion years old (Nutman et al. 2016). The sophistication of microorganisms in these stromatolites raises questions. One possibility is that life originated much earlier, but records in the rocks have been lost. Another possibility is that perhaps the stromatolites might have been photosynthetic but not the oxygen-liberating type of photosynthesis of modern cyanobacteria. However, it should be noted that banded iron deposits are thought to form in conjunction with the liberation of iron by cyanobacteria, and there are 3.7-billion-year-old banded iron deposits (Rosing et al. 1996). But, from 3.5 billion years ago to the present, there are well-documented examples of stromatolites. Perhaps it is at least safe to say that the filamentous bacteria-like structures in these formations show that they contain carbon and the ^{12}C to ^{13}C ratios correspond to different types of bacteria, including various members of the Archaea domain (Schopf et al. 2018).

There is little in the paleontology that shows us how life arose. If anything, the data we do have pushes the origin of life into a period where we have no record at all. What we have been able to do is identify the earliest records for life on earth and establish that early life consisted of microbes. However, there are other ways to approach answers to this question, one of which is the search for the last universal common ancestor (LUCA) of living things.

© Marco Tomasini/Shutterstock.com.

Figure 1.3. Modern stromatolites in Shark Bay, Western Australia. These stromatolites are only a few thousand years old but they are similar to those of the ancient past.

3 The Search for the LUCA Using DNA/RNA Sequencing

There is a branch of biology known as molecular phylogenetics that uses DNA or RNA sequencing to identify relationships between organisms. The field got its start from work done by Zuckerkandl and Pauling (1965a, b), who were comparing the amino acid sequences of hemoglobin from different organisms. They proposed the idea that mutations cause changes in the amino acid sequence to occur at a roughly constant rate, giving rise to the idea of an evolutionary molecular clock, where the amino acid differences in the hemoglobin of different organisms were proportional to the time from their divergence from a common ancestor.

If there has been an unbroken chain of living things from the first living organism to those of present day, we might be able to trace the history of organisms by protein, DNA, or RNA sequences. However, one must make some assumptions: for example, common ancestry. There are many lines of evidence that suggest that there is a common ancestry to living things. These include the facts that all organisms use the same mirror image forms (called enantiomers) of amino acids and sugars, that heredity is based on nucleic acids, that the genetic code is nearly universal, and that we share the same basic metabolic pathways. A recent study of thirteen proteins shared by all organisms tested which model best fits the data, based on a universal ancestor or independent origins, and the results overwhelmingly supported the common ancestry model (Theobald 2010).

A. DNA Preserves the History of Mutations

Mutations are changes in the sequence of bases in DNA. Mutations may interfere with functions of organisms. Some mutations are in parts of the DNA that do not code for proteins. Some mutations change a base in the DNA, but the changed nucleotide still ends up coding for the same amino acid. Some mutations do change the amino acid, but the change hardly affects the function of a protein. For example, the change from leucine to isoleucine often has little effect because both are very similar hydrophobic amino acids. Finally, some mutations damage proteins. However, for analysis of phylogenetic relationships between organisms, any change is a useful marker.

The simplest change, then, is the change of one DNA or RNA base. We as humans carry many differences in our genes where one base is different from one person to another. This is known as single nucleotide polymorphism.

1) AAGCCTA to AAGCTTA

At present there are about 1.5 million known single nucleotide polymorphisms points in the genome where human genomes differ from one another. As we move from comparing genomes from different humans to comparing humans to other organisms, we see more and more differences in genomes. Generally, genes that code for proteins or RNA change less than noncoding DNA. Ribosomal RNA is particularly useful because all organisms – both prokaryotes and eukaryotes – have ribosomal RNA.

In the simplest application, one compares the bases that differ. Then one can divide that number by the total number of bases compared to get the percentage of divergence.

B. Problems with Trying to Reconstruct Relationships Using DNA or RNA Sequences

1) Lateral Gene Transfer

In fact, DNA can be inherited laterally (between two unrelated organisms) – which is referred to as lateral gene transfer. Some think that the problem of lateral gene transfer is so great that it has made it impossible to trace genes back to the LUCA, (Doolittle, 1999).

a. Syncytin-A and -B as Examples of Lateral Gene Transfer

The human placenta contains several cellular layers that separate the uterus from the fetus. One of these layers is the syncytiotrophoblast layer.

A syncytium is formed when individual cells fuse together to form one giant multinuclear cell. The proteins that cause cells of the syncytiotrophoblast layer to fuse together are syncytin-A and -B.

When the genes for syncytin-A and -B were analyzed, something remarkable was found. Both syncytin-A and -B genes were part of what appeared to be canonical retroviral genomes. The long terminal repeat (LTR) is how the retroviral genes integrate into the genome. The Pol gene codes for reverse transcriptase and integrase, the enzymes that copies retroviral RNA into DNA and inserts it into the genome. The Gag protein makes up the capsid of the virus, and the Env gene is a protein in the membrane envelope that surrounds the virus.

You can learn more about retroviruses here: https://www.youtube.com/watch?v=dn1tNIrMPRk

It turns out that syncytin-A and -B are the envelope proteins of retroviruses that infected and integrated into our genomes. The other retroviral (Pol and Gag) genes have been inactivated. This is an example of lateral gene transfer.

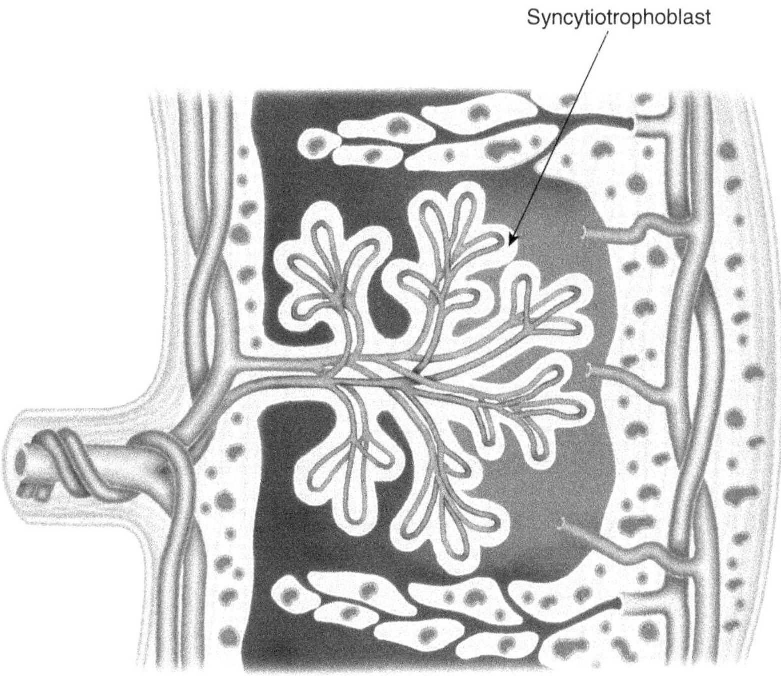

Figure 1.4. Human placenta cellular layer.

2) Long Branch Attraction

Long branch attraction is a complicated problem that can apply to both character traits or to DNA/RNA sequences. For our purposes, it is the false appearance of relatedness of two long branches in a sequence comparison. In a long branch attraction, whether due to DNA sequences evolving rapidly or simply due to the length of the branch, multiple mutations can occur at sites along the DNA being compared. This can cause DNA sequences to appear more similar than their true divergence or relatedness. For example, What if an A originally mutated to a T but then with time, that same nucleotide mutated again, now to a G and then to a C and finally back to an A? It would look like nothing ever happened, when in fact, a lot happened. This can make two organisms appear to be more closely related than they really are, leading to serious problems in attempting to create a phylogenetic tree.

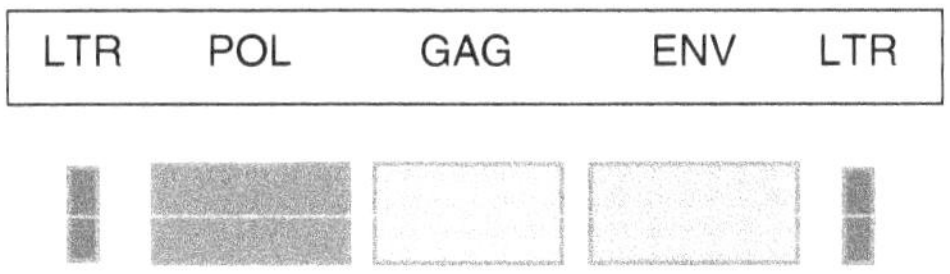

Figure 1.5. Retroviral genome.

Source: M. Poenie.

C. Ribosomal RNA Analysis Reveals Three Domains of Life

Despite these problems, there has been some progress. A major step forward was made by Woese and Fox (1977) when they made sequence comparisons of ribosomal RNA from a broad swath of organisms. They were initially confounded because some of their data was not making sense. Certain bacteria-like organisms appeared to be "bacterial freaks," and Woese commented it was like they were looking at something from Mars. When Woese and Fox finally published their findings, they identified a new form of life, known as the archaea, that looked like bacteria but were remarkably different. Furthermore, many of these archaea live in extreme environments such as hot springs or under anaerobic conditions, and it was thought that they may represent the oldest living on earth.

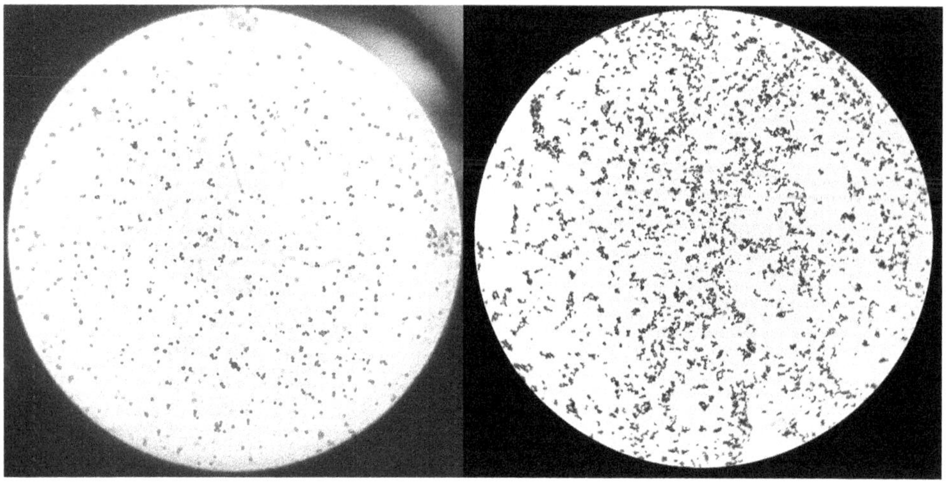

Figure 1.6. Microscopic images of archaea (left) and eubacteria (right).

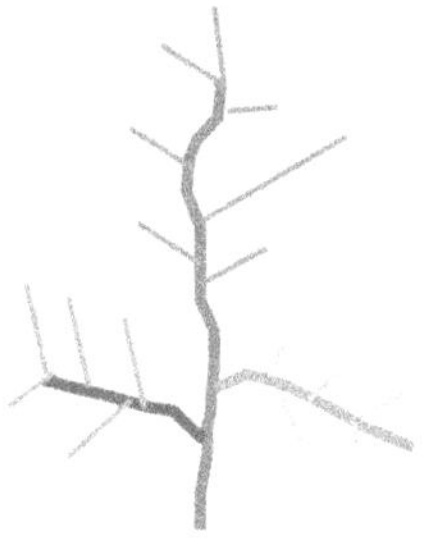

Figure 1.7. Eukarayotes.

Source: M. Poenie.

Both archaea and true bacteria (eubacteria) are prokaryotes and look quite similar to each other on the surface, but genetically they are quite different. The archaea share many similarities to the eukaryotes in their DNA that codes for information-related functions (e.g., enzymes participating in transcription, translation). On the other hand, eubacteria and archea are similar in their DNA for their metabolic enzymes.

Before the work of Woese and Fox, life was basically divided into plant and animal kingdoms, although some classification systems were more detailed. Now, life would be classified in terms of three domains: Eubacteria, Archaea, and Eukarya. It is possible that this three-domain classification might revert back to two, because numerous genes that were once considered signature eukaryotic genes have been now been identified in the archaea. In other words, eukaryotes might be more closely related to the archaea than the division into three domains suggests.

4 Prokaryotes

Both eubacteria and archaea are prokaryotes which have a much simpler internal structure than the eukaryotes. Prokaryotes do not have the internal membrane-enclosed compartments (organelles) that are typical of eukaryotes. The DNA of prokaryotes primarily consists of a single circular chromosome, although they can carry additional small circular pieces of DNA known as plasmids.

While the archaea are superficially similar to bacteria, they exhibit remarkable and fundamental differences from the eubacteria in their biochemistry. For example, the cell membrane lipids are different. In both eubacteria and eukaryotes, membrane lipids are made from one of the two mirror-image forms of glycerol (D), whereas archaea use the other isoform (L). Instead of the typical fatty acids used as membrane lipids in the eubacteria and eukaryotes, archaea use branched chain lipids called prenyl lipids because they are derived from isoprene. Furthermore, these lipids

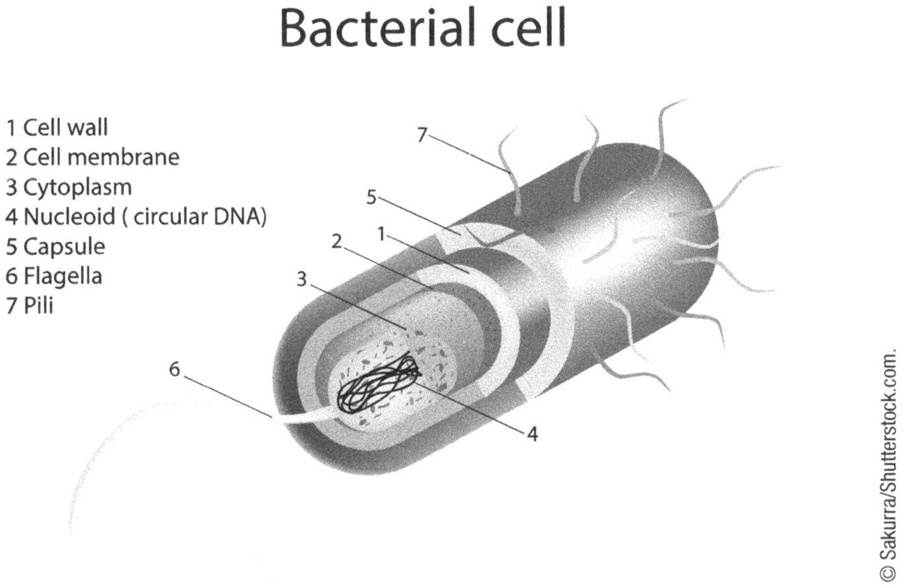

Figure 1.8. Bacteria. Prokaryotes are structurally simple compared to eukaryotes. As a rule they do not have the numerous membrane enclosed compartments typical of eukaryotic cells.

are linked to glycerol through ether linkages rather than the ester linkages used to link fatty acids to glycerol in bacteria and eukaryotes. The cell wall is also very different in its composition from that of the eubacteria. Therefore, antibiotics that disrupt the cell wall of eubacteria, such as the penicillin family, have no effect on archaea.

While some features of archaea are different from both bacteria and eukaryotes, some aspects of the archaea are much more similar to eukaryotes than they are to bacteria. In particular, archaea and eukaryotes are closer to each other when it comes to transcription and translation and DNA replication.

Examples of these differences are seen in different sensitivities to antibiotics and inhibitors. For example, the ribosomes of eubacteria and eukaryotes are different in both size and antibiotic sensitivity. Streptomycin and diphtheria toxin block the bacterial ribosome but not those of the eukaryotes or archaea. Similarly, anisomycin, which blocks the eukaryotic ribosome, also

Table 1.1. A Number of Archaean Genes, Especially Those Involved in Information Processing, Are More Similar to Eukaryotic Genes than to Bacteria

	Bacteria	Archaea	Eukaryotes
Sensitivity to streptomycin (blocks bacterial ribosome)	Yes	No	No
Sensitivity to anisomycin (blocks eukaryotic ribosome)	No	Yes	Yes
Sensitivity to diptheria toxin	No	Yes	Yes
Sensitivity to aphidicolin (DNA polymerase inhibitor)	No	Yes	Yes
Histones on DNA	No	Yes	Yes
DNA replication requires CDC6, MCM	No	Yes	Yes
Proteins begin with	fMET	MET	MET

blocks that of archaea but not bacterial ribosomes. Aphidicolin, an inhibitor of eukaryotic DNA polymerases, also blocks the polymerase of archaea but not that of eubacteria. In eukaryotes, DNA is wrapped around histone proteins in structures known as nucleosomes. Archaea but not bacteria also have histone proteins that serve to organize the DNA. In eukaryotes, the start of DNA synthesis requires the loading of a DNA unwinding helicase, known as MCM, onto the DNA. Loading of MCM requires an accessory protein known as CDC6. Archaean DNA synthesis, in contrast to bacteria synthesis, requires proteins similar to MCM and CDC6. Finally, the synthesis of bacteria proteins starts with formylmethionine amino acid, whereas in eukaryotes and archaea, the first amino acid is methionine. So in many ways the archaea are more similar to eukaryotes than to bacteria.

A. Signature Eukaryotic Proteins Are Discovered in a Branch of the Archaea

For a long time, obtaining DNA sequencing of prokaryotes required that particular organisms be grown in the lab. This limited the species that could be analyzed. Recently, it has become possible to obtain genome sequences from a wider range of organisms. New sequence data have revealed even more stunning similarities between archaea and eukaryotes in the form of eukaryotic signature proteins. Eukaryotic signature proteins were originally defined as those proteins found in all branches of the eukaryotes but not in the eubacteria or archaea. However, recent studies have found that certain groups of the archaea known as the TACK superphylum contain many eukaryotic signature genes. In particular, the Lokiarchaeota have over 100 signature eukaryotic proteins. This new evidence suggests a close relationship between the archaea and eukaryotes.

B. Examples of Signature Eukaryotic Proteins Found in Archaea

A. Actin and the Arp 2/3 (actin-related proteins)

B. Ribosomal proteins S25e, S30e, L13e and L38ev

C. The eukaryotic RNA polymerase III subunit Rpc34

D. The eukaryotic transcription elongation factor Elf1

E. Proteins of the ESCRT membrane remodeling machinery

F. A cluster of genes that correspond to the ubiquitin modifier enzymes

G. Artubulin – what appears to be a true ancestral tubulin

5 Eukaryotes

The eukaryotes are radically different in structure from both archaea and eubacteria. Eukaryotes are full of organized structures, many of which are enclosed in membranes, that carry out specific cellular processes. Membrane-enclosed structures include the nucleus, mitochondria, lysosomes, endoplasmic reticulum, peroxisomes, and the Golgi apparatus. Nonmembrane-enclosed structures include the centrioles, which are part of the centrosome, and the nucleolus, which is the site in the nucleus where ribosomal RNA is synthesized.

A. Eukaryotic Cell Structure

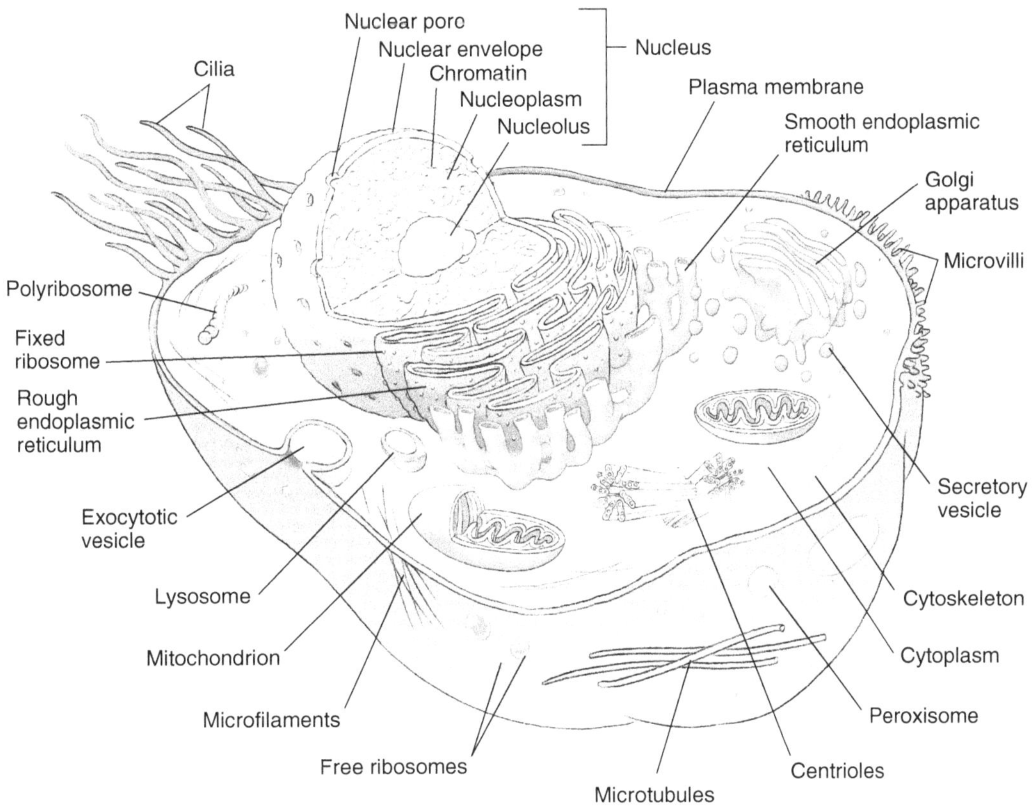

Figure 1.9. The eukaryotic cell exhibits a much more complicated internal structure than the prokaryotic cell.

B. Origin of Eukaryotes and Their Organelles

1) The First Eukaryotes

The origin of eukaryotes, like that of the first cells, is shrouded in mystery. There really are no modern representatives of what we might call a "primitive" eukaryote. From our vantage point, it seems that the last common ancestor of eukaryotes had all the same organelles as modern eukaryotes. The one thing we can say with some confidence is that modern eukaryotes represent a combination of an archaean and a eubacterium. This idea was popularized by Lynn Margulis (Sagan 1967; Margulis and Bermudes 1985), who suggested that the mitochondria and chloroplasts of eukaryotes were originally bacteria that came to be symbiotically associated with the eukaryotic cell.

2) The Origin of Eukaryotes through Symbiosis

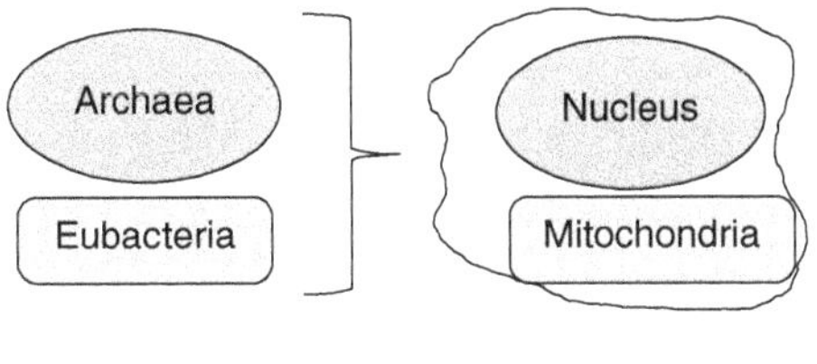

Figure 1.10.

There are many variations on the theme of the endosymbiosis theory. One hypothesis is that a eubacterium known as an alpha-proteobacterium engulfed an archaean cell, ultimately giving rise to a cell where the bacterium became the mitochondria and the archaean became the nucleus. At the other end of the spectrum, it has been suggested that a protoeukaryote capable of crawling around emerged from the archaea and then at some point, this protoeukaryote engulfed a eubacterium, which, instead of being digested, eventually became the mitochondria.

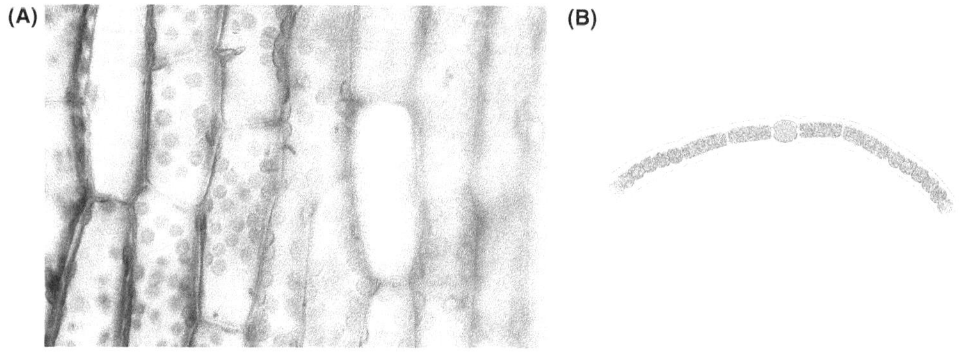

Figure 1.11. **(A)** The image shows the green chloroplasts inside a plant cell **(B)** An image of photosynthetic cyanobacteria.

Source: A: Rattiya Thongdumhyu/Shutterstock.com. B: Aldona Griskeviciene/Shutterstock.com

3) Similarities between Mitochondria, Chloroplasts, and Prokaryotes

There are many similarities between eubacteria and the mitochondria on the one hand and between bacteria and chloroplasts on the other. Mitochondria and chloroplasts are self-replicating organelles that, like bacteria, have their own DNA, which they transcribe to mRNA and which is then translated to make proteins. Euacteria have 70S* ribosomes whereas eukaryotic cytoplasmic ribosomes are 80S. However, mitochondria and chloroplasts have their own ribosomes that are 70S like bacteria. The first or initiator amino acid of eubacterial proteins is fMET, but the cytoplasmic proteins of eukaryotes start with MET. Given the similarities between mitochondria and eubacteria, what amino acid (fMET or MET) would you expect to initiate protein synthesis in mitochondria and chloroplasts?

Table 1.2. Comparison of prokaryotes, eukaryotes, mitochondria and chloroplasts

	Prokaryotes	**Eukaryotes**	**Mitochondria**	**Chloroplasts**
DNA/Chromosomes	1 circular	Multiple linear	1 circular	1 circular
Replication	binary fission	mitosis	binary fission	binary fission
Ribosomes	70S	80S	70S	70S
Electron transport	plasma membrane	No	inner membrane	thykaloid membrane
Initiator amino acid	fMET	MET	fMET	fMET

*S refers to Svedburg units – a measure of the rate that things sediment when placed in a centrifuge. So basically it is a measure of how big and heavy molecules or structures are.

C. Symbiogenetic Genes

An additional line of evidence for the symbiotic origin of mitochondria and chloroplasts is seen in the nuclear DNA of eukaryotes. While mitochondria can make their own proteins, in humans they make only about 37 proteins, compared to the thousands of proteins made by eubacteria. So one wonders, "What happened to all the other genes that were once present in the eukaryote?" It turns out that many of them were transferred to the nucleus of the eukaryotic cells. We see evidence that many eukaryotic genes are actually fusions of a eukaryotic gene with fragments of a bacterial gene. These fusions have been called symbiogenetic genes or S genes (Meheust et al. 2018).

While there has been progress in understanding the origin of cells, we are still far from understanding how life got started on earth. The evidence indicates that prokaryotes are at the base of the tree of life and that eukaryotes represent a combination of elements of both eubacteria and archaea, although they are more closely related to the archaea. Photosynthetic prokaryotes appear to be some of the earliest living things based on the records we have, although the photosynthetic systems used by these organisms is uncertain.

6 Fractionation and Purification of Organelles from Eukaryotic Cells

A. Pioneering Efforts to Understand Structure and Function of Organelles

There are many different organelles that distinguish eukaryotes from prokaryotes. Here we begin to address the question of how we know what these different organelles do. One of the early pioneers of cell biology was Christian De Duve, who discovered the lysosome and peroxisome organelles. He received the Nobel Prize in 1974 along with Albert Claude and George Palade, who were also important pioneers in understanding the function and organization of the cell. Interestingly, De Duve discovered lysosomes without having a microscope in his laboratory (Sabatini and Adesnik 2013). Instead, he pioneered the use of centrifugation to isolate and purify organelles. Today, the centrifuge is a valuable tool for cell biology. Below, we discuss how it is used.

B. Separation of Organelles by Centrifugation

In the purification of organelles, we first need a way to open up cells and release the organelles without damaging them, which is referred to as homogenization. There are many ways to do this, but often it involves squeezing cells through a tight space under pressure. Examples shown below are the Dounce homogenizer, a glass tube that holds the cells and the media and a glass plunger that breaks the cells as they pass between plunger and the tube. Alternatively, syringes can be used. In this case cells in media are drawn into the syringe and then forced several times under pressure through a narrow needle.

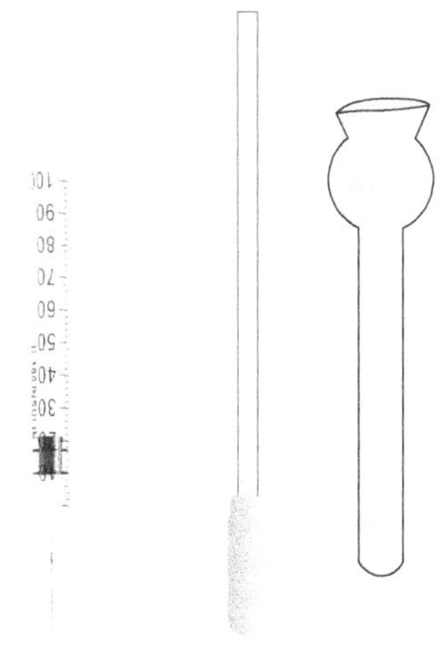

Figure 1.12. Examples of ways to homogenize cells. Left, by forcing a cell suspension through a syringe needle. Right, a Dounce type homogenizer where the plunger is forced down the tube containing the cell suspension.

Syringe: Davydenko Yuliia/Shutterstock.com. Dounce type homogenizer: © Kendall Hunt Publishing Company.

Once one obtains a cell homogenate, the organelles can be isolated using a variety

© Anucha Cheechang/Shutterstock.com.

Figure 1.13. Photo of the inside of a centrifuge with a lab worker having one hand on the rotor and the other hand holding the rotor lid. Centrifuges often need to be refrigerated, and some centrifuges spin so fast that the chamber must be kept under high vacuum to prevent rotor heating.

of centrifugation techniques. These include differential, rate zonal, and isopycnic centrifugation. In each case, the centrifugation technique involves transferring the homogenate to a centrifuge tube and then placing the tube in the rotor of the centrifuge. The rotor is then rotated at a speed that depends on the particular isolation protocol. As the rotor is spun, a centrifugal force develops that pushes cell organelles from the top of the tube toward the bottom. This force is usually quantified in terms of how strong it is relative to gravity (relative centrifugal force, or RCF) so the force generated by the centrifuge is expressed as × g (times gravity). This is a useful measure because the force generated by the centrifuge depends on both the speed of rotation and the radius of the rotor. By expressing centrifugal force as × g, both factors are taken into account.

C. Differential Centrifugation

Differential centrifugation separates organelles on the principle that some organelles will pellet at the bottom of the tube at relatively slow speeds whereas for other organelles, it takes much higher speeds. For example, the nucleus of cells is relatively large and heavy, and it will spin to the bottom of the tube at 600 × g after only 10 minutes. Once the nuclei have formed a pellet at the bottom of the tube, the remainder of the homogenate (the supernatant) can be poured off or poured into a second tube. This second tube receives a homogenate that has everything in it except nuclei. This second tube is now put in the centrifuge and subjected to 15,000 × g for 5 minutes. This gives a new pellet consisting of mitochondria, chloroplasts, peroxisomes, and lysosomes (depending on cell type). The supernatant is poured off into a third tube, and this tube subjected to 100,000 × g for an hour, which pellets various membrane fragments referred to as microsomes and poly ribosomes. Finally the supernatant from the third tube is poured into a fourth tube and this is subjected to 300,000 × g for 3 hours. This final spin pellets free ribosomes.

It should be noted that the actual g forces might vary somewhat since pelleting of organelles depends both on the g force and the time of the centrifugation.

D. Rate Zonal and Isopycnic Centrifugation

1) Making Gradients

Both rate zonal and isopycnic centrifugation depend on preparing centrifuge tubes filled with a material whose density varies from the bottom of the tube (greatest density) to the top of the tube (lowest density). One of the most common materials for making gradients is sucrose. The density of 1 molar (M) sucrose is 1.127 grams/ml and the density of 2 molar sucrose is 1.253 grams/ml. Sucrose gradients are easily prepared using a gradient maker. The gradient maker is a device with two chambers, the mixing chamber and the reservoir. One usually chooses the range of sucrose concentrations that will best separate the organelle of interest. The more dense sucrose solution is placed in the mixing chamber and the lighter sucrose solution placed in the reservoir. The two valves prevent mixing until the gradient formation starts. A magnetic stir bar is placed in the mixing chamber and the device is placed on a magnetic stirrer. The valves are opened, and the initial solution coming out of the maker will be the most dense. However, as the flow progresses, the lighter solution in the reservoir will flow into the mixing chamber and the solutions will be mixed by the spinning stir bar. The mixing in of the lower density sucrose gradually makes the solution flowing out of the gradient maker progressively less dense, with the lowest density sucrose at the top of the tube.

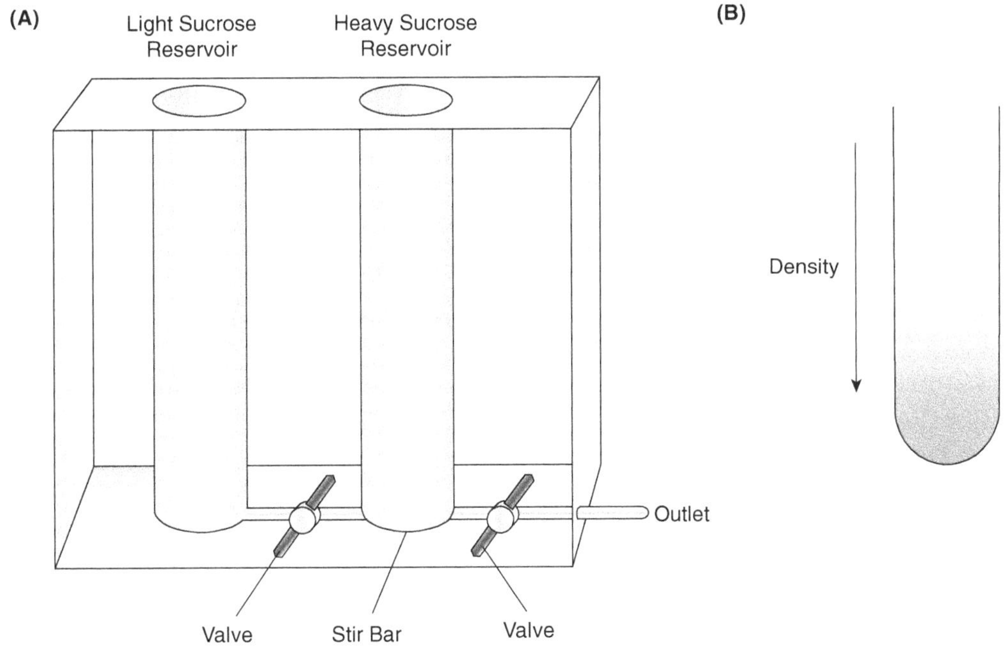

Figure 1.14. (A) A common gradient maker **(B)** Illustration of a sucrose gradient in a centrifuge tube where darkness of color corresponds to the density of sucrose.

A: © Kendall Hunt Publishing Company. B: Source: M. Poenie.

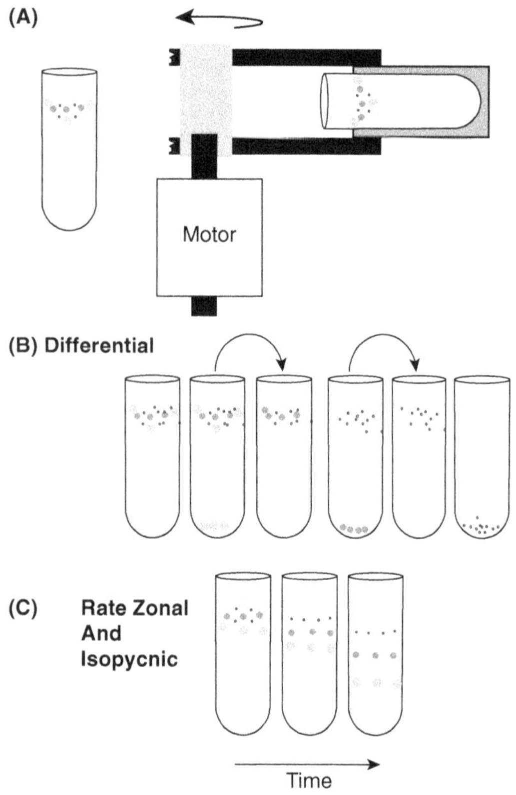

Figure 1.15. Centrifugation. **(A)** illustrates the homogenate, which is a mixture of differently sized organelles. **(B)** Differential centrifugation: The homogenate is first subjected to low-speed centrifugation, which pellets the largest organelles. Subsequent centrifugation at increasingly higher speeds and/or for longer times pellets smaller organelles. **(C)** Rate Zonal and Isopycnic centrifugation: In this case the homogenate is subjected to centrifugation through a gradient of increasing density, typically through different concentrations of sucrose. In rate zonal centrifugation, the organelles move at different rates through the sucrose gradient but they never stop moving. In isopycnic centrifugation, organelles move to a point in the gradient where their density matches that of the sucrose and then they stop.

2) Rate Zonal Centrifugation

In rate zonal centrifugation organelles separate out by the rate they move through the centrifuge tube. The rate they sediment is a combined function of density, size, and shape of the particle as well as the viscosity of the medium and the force applied. We will not be calculating sedimentation rates, but rates are given in terms of Svedburg units (S).

a. Isopycnic-Equilibrium Density Gradient Centrifugation

In Isopycnic centrifugation, separation of organelles does not depend on the rate they move through the gradient. The sucrose range between the top and bottom is chosen such that the density at top of the tube is less than that of the organelles whereas the sucrose at the bottom of the tube is more dense than the organelles. Under these conditions, the organelles will move through the sucrose until the density of the organelle matches that of the sucrose. At this point, no further movement takes place.

Christian Deduve's Nobel prize lecture was entitled, "Exploring cells with a centrifuge." Since his pioneering studies, the centrifuge has been used for everything from separating cells to separating assemblies of molecules. It is one of the extremely important complement of tools that cell biologists use. By purifying organelles, it has been possible to study their properties, their composition, and how they work.

References

Doolittle, W.F., *Phylogenetic classification and the universal tree*. Science, 1999. **284**(5423): p. 2124–9.

Margulis, L., and D. Bermudes 1985. "Symbiosis as a Mechanism of Evolution: Status of Cell Symbiosis Theory." *Symbiosis* 1: 101–24.

Méheust, Raphaël, Debashish Bhattacharya, Jananan S. Pathmanathan, James O. McInerney, Philippe Lopez, and Eric Bapteste. 2018. "Formation of Chimeric Genes with Essential Functions at the Origin of Eukaryotes." *BMC Biology* 16 (1): 30.

Mojzsis, Stephen J., T. Mark Harrison, and Robert T. Pidgeon. 2001. "Oxygen-Isotope Evidence from Ancient Zircons for Liquid Water at the Earth's Surface 4,300 Myr Ago." *Nature* 409: 178–81.

Nutman, A. P., V. C. Bennett, C. R. Friend, M. J. Van Kranendonk, and A. R. Chivasl. 2016. "Rapid Emergence of Life Shown by Discovery of 3,700-Million-Year-Old Microbial Structures." *Nature* 537 (7621): 535–38.

Sagan, Lynn. 1967. "On the Origin of Mitosing Cells." *Journal of Theoretical Biology* 14 (3): 225–74.

Theobald, D. L. 2010. "A Formal Test of the Theory of Universal Common Ancestry." *Nature* 465 (7295): 219–22.

Valley, John W., William H. Peck, and Elizabeth M. King. 2002. "A Cool Early Earth." *Geology* 30 (4): 351–54.

Wilde, Simon A., John W. Valley, William H. Peck, and Colin M. Graham. 2001. "Evidence from Detrital Zircons for the Existence of Continental Crust and Oceans on the Earth 4.4 Gyr Ago. *Nature* 409: 175–78.

Woese, C. R., and G. E. Fox. 1977. "Phylogenetic Structure of the Prokaryotic Domain: The Primary Kingdoms." *Proceedings of the National Academy of Sciences of the United States of America* 74: 5088–90.

Zuckerkandl, Emile, and Linus Pauling. 1965a. "Evolutionary Divergence and Convergence in Proteins." In *Evolving Genes and Proteins*, edited by V. Bryson and H. J. Vogel, 97–166. New York: Academic Press.

______. 1965b. "Molecules as Documents of Evolutionary History" *Journal of Theoretical Biology* 8: 357–66.

Microscopy

2

The microscope is one of the cell biologist's most important tools. It is what mainly distinguishes cell biologists from molecular biologists. Cell biology is defined as the function of cells in terms of their component molecules. Cellular functions are subdivided into compartments called organelles, where different organelles carry out specific and unique functions (Figure 2.1). For example, a cell has a nucleus which contains DNA, the hereditary information of the cell. Within the nucleus is a region of intense RNA synthesis called the nucleolus devoted to making ribosomes. The nucleous manufactures ribosomal RNA (rRNA). Ribosomes are structures made of rRNA and protein that are used to manufacture new proteins. In the nucleus, information or instructions for making proteins is copied from DNA into an RNA molecule in a process called transcription. Once processing is finished, this RNA is carried into the cytoplasm and used by ribosomes to make a specific protein in a process called translation.

The cell contains many other organelles. For example, there are mitochondria that make ATP, the main energy source for driving cell functions. There are lysosomes, which are acidic inside and are

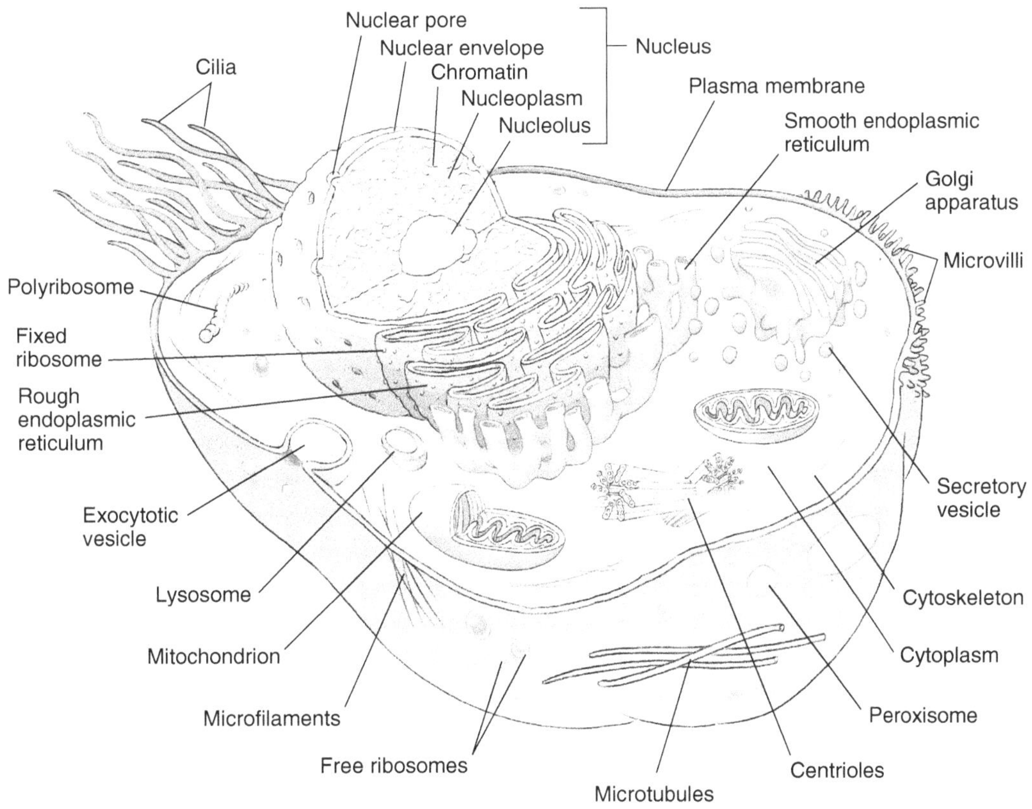

Figure 2.1. Diagram of a cell and its organelles.

used to digest incoming food molecules such as proteins, lipids, or nucleic acids. These organelles are too small to see with our naked eye and some are either too small or too transparent to see using a typical light microscope. However, cell biologists have many tools that, combined with different types of microscopes, help them to see cell structure.

In discussing cell biology we need a common system of measurement to describe the dimensions of cells and their subcellular components. We use the metric system, which divides up a meter into subdivisions of tenths, hundredths, thousandths, and so on. Sizes of different familiar objects are shown in Figure 2.2. A millimeter is a thousandth of a meter and, while a few types of cells such as frog eggs can be more than a millimeter in diameter, most cells are much smaller than a millimeter. Below the millimeter, the next subdivision is micrometers or millionths of a meter (μm or 10^{-6} meter)

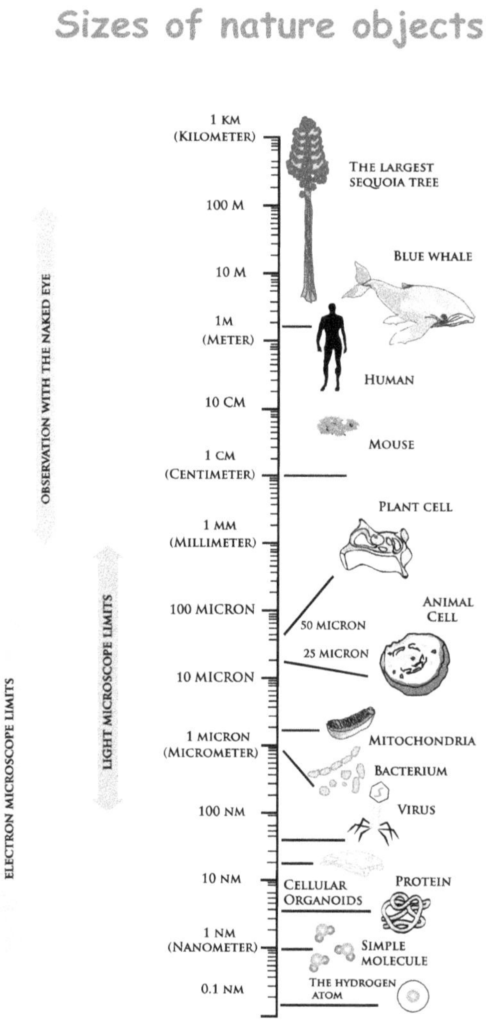

Figure 2.2. Sizes of things we measure using the metric system. Millimeters are 10^{-3} meters. Micrometers (microns) refer to 10^{-6} meters. Nanometers refer to 10^{-9} meters, and angstroms refer to 10^{-10} meters or 0.1 nanometers. Cell diameters and some organelles are measured in microns whereas some cell components are measured in nanometers.

and often referred to as microns. The diameter of most cells is typically given in microns. Some parts of a cell such as the nucleus or mitochondria can be measured in microns but many components are much smaller than one micron, so we need to go to the next subdivision, billionths of a meter, referred to as nanometers (nm or 10^{-9} meter). Most of the components of a cell, including individual protein molecules, can be measured in nanometers but when it comes to atoms we need yet a smaller division, one tenth of a nanometer, which is the angstrom (Å or 10^{-10} meter).

In order to see things as small as the cell and its component parts, we need to use a microscope. Microscopes are important for their ability to magnify small objects, but we need more than just magnification. Key features of a microscope are its *resolution* and *contrast*. Resolution is defined as a distance how far apart two small objects must be in order to tell that they are separate objects. We will see that light microscopes have limits on how well they can resolve small objects that are close together. At some point, as two small objects are brought closer and closer together, the images of these objects will merge and one can no longer distinguish them as separate. At some point, due to the limited resolution, it will not matter how much you magnify the image. Further magnification will magnify the blurred image of the two objects but it will not reveal that they are really separate objects. The resolution we can obtain from the light microscope is limited by the properties of light, which we will discuss in Section 1.

Contrast is a separate issue that limits what we can see under the microscope. Cells consist mostly of water. So, when we look at a cell in its normal watery environment it is almost (not quite) as if we are looking at a drop of water in water. It is a challenge to distinguish the cell and its component parts from the background. The issue of contrast is a difficult problem but microscope techniques have been developed to improve contrast.

1 The Physics of Light

We will start our venture into microscopy by discussing light microscopes that operate in the visible part of the electromagnetic spectrum. Light is electromagnetic radiation that can travel through empty space and that has characteristics of both waves and particles (*photons*). As a wave, light oscillates and can be characterized by its wavelength, the distance from one wave peak to the next peak (Figure 2.3). The entire electromagnetic spectrum includes radiation of wavelengths that vary from picometers (10^{-12} meters) known as gamma rays to radio waves whose wavelengths extend to kilometers (10^3 meters). Visible light represents a very narrow slice of the electromagnetic spectrum covering wavelengths from about 400 to 700 nm (nm = nanometers or 10^{-9} meters).

Light waves can also be viewed as a stream of photons where each photon is a packet of energy. The energy in a photon is related to the wavelength of the light such that for our discussion, energy and wavelength are interchangeable ideas. The shorter the wavelength, the higher the energy of the photon. The concept of a photon is especially important when we consider light interacting with devices like cameras or molecules that absorb light.

A. Visible Light

When you see a rainbow, you are seeing light of different wavelengths separated out into different bands (a spectrum, Figure 2.4) due to the phenomenon of refraction. On one side of the rainbow is red, corresponding to the long wavelength end of the visible light spectrum. On the opposite side

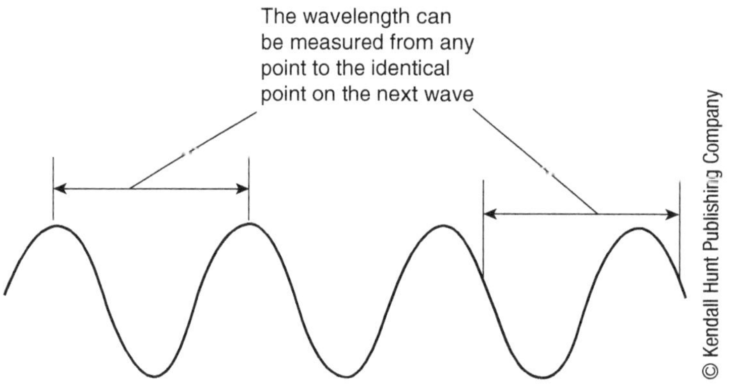

Figure 2.3. Electromagnetic radiation is a form of energy that travels through space as a wave. Electromagnetic waves are characterized by their wavelength, the distance from one crest to the next. Electromagnetic wavelengths vary widely, from gamma rays at the short end with wavelengths measured in picometers (10^{-12} meters) to radio waves which have wavelengths that range from millimeters to kilometers. Sandwiched in between these two extremes is visible light, which covers wavelengths from 400 to 700 nanometers (10^{-9} meters).

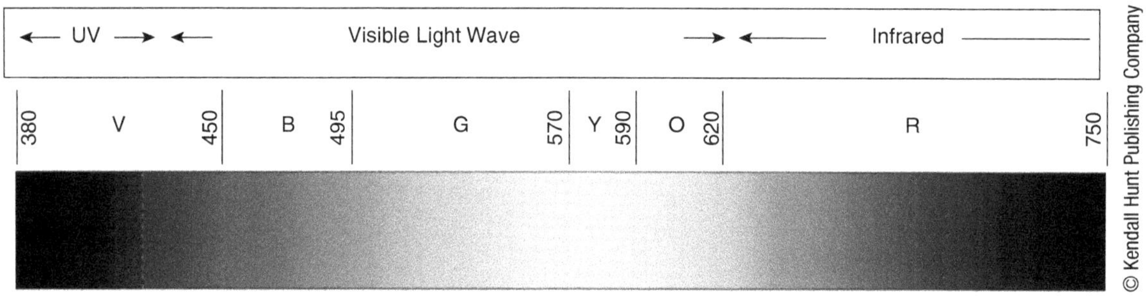

Figure 2.4. Visible light refers to electromagnetic radiation with wavelengths between about 400 and 700 nm. Our eyes detect different wavelengths of light as different colors, with the shortest wavelength light appearing blue and the longest as red. For radiation going in the progressively shorter wavelength direction (going left of blue), we go progressively through ultraviolet and then X-rays and finally to gamma rays. On the opposite side of the spectrum, wavelengths are getting progressively longer from red to infrared to microwaves and finally to radio waves. We can see only a very narrow band of the whole spectrum.

is purple and blue, corresponding to the short wavelength end of the visible spectrum. This is due to the fact that different wavelengths of light are refracted differently by the droplets of water that cause the rainbow to form.

Light can interact with matter in a number of ways, but these interactions often depend on the wavelength of the light. One type of interaction is absorption, where the energy of a photon of light can be transferred to atoms or molecules. What happens at that point depends on the behavior of the molecule or atom. The absorbed energy of a photon must go somewhere but there are a number of possible things that can happen, including conversion to heat, altering chemical bonds, or the reemission of the energy as a new photon.

We see different colors because different photoreceptors called cone cells in our eyes have pigments that absorb different wavelengths of light (Figure 2.5). There are red, green, and blue cone cells that each preferentially absorb light in a different region of the spectrum. There is a fourth type of photoreceptor known as rod cells that preferentially absorb in the green region of the spectrum, but this is interpreted as different shades of gray.

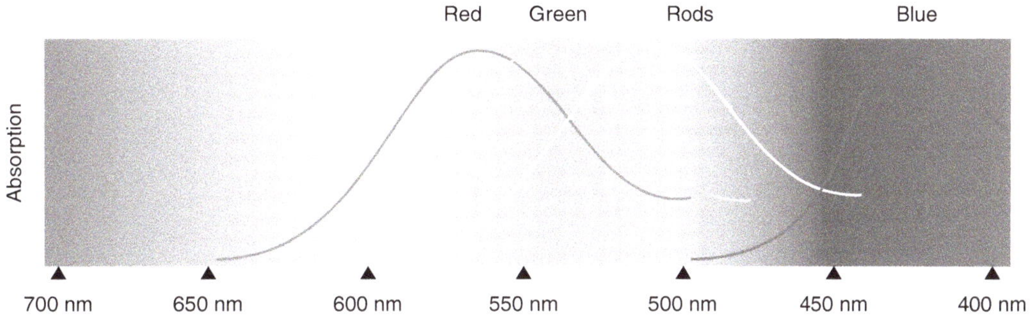

Figure 2.5. Absorbance spectra for the blue, green, and red cone cells responsible for our color vision. The absorbance spectrum of the rod pigment is also shown. Each type of cone cell exhibits a peak in its absorbance at different points in the spectrum, which in turn determines our perception of color.

B. Diffraction

There are interactions of light with objects that do not involve absorption. One of these interactions is called diffraction. Diffraction refers to the way light bends or spreads out when passing small openings or small objects. It is a basic property of waves, including waves of water (Figure 2.6). Diffraction is readily observed by passing light through a narrow slit. As light passes through the slit it spreads out, and if the light is now projected onto a screen or wall, one will see a series of bright spots or bands. The bright and dark bands are referred to as a diffraction pattern, which arises due to the phenomenon known as interference. Interference (see Section 1.C) occurs because waves arising from different regions of the slit travel different distances to the screen and therefore arrive at particular points on the screen out of phase from each other.

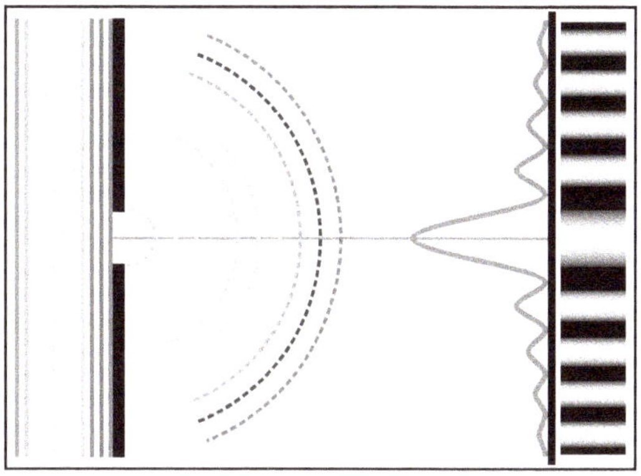

Figure 2.6. Left diffraction of water waves. Right diffraction of light giving rise to a diffraction pattern.

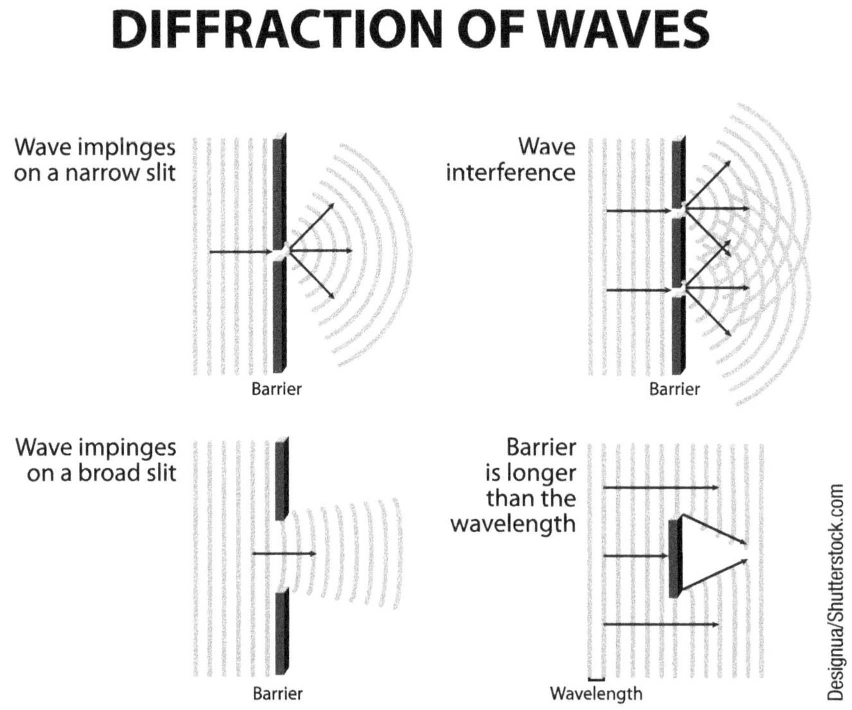

Figure 2.7. Diffraction and the dimensions of the opening or object. If light passes through a small slit, it spreads out more than if it passes through a wider opening. The key is the relationship between the size of the opening or object relative to the wavelength of the light. For any given size of slit, short wavelength light will spread less than long wavelength light.

C. Interference

Interference refers to how two or more light waves crossing the same point in space interact with each other. If the crests of two waves coincide in a region of space, the result is a wave with a higher crest. This is known as constructive interference and where this takes place, the area will appear brighter. On the other hand, if the crest of one wave coincides with the valley of another wave, the two waves cancel each other. This is known as destructive interference and where this takes place, the area will appear dark. It is perhaps easiest to see the phenomenon of interference by have light pass through two slits at the same time.

When light passes through two slits, it will spread out from both slits due to diffraction. The reason that light forms patterns of constructive and destructive interference is that waves of light coming from the two different slits intersect, having traveled different distances to reach the same point on the screen Figure 2.8. In the center, all light rays are taking essentially the same path and the result is constructive interference, giving rise to the bright central spot. At points to the left or right of the central spot the distances are not the same. This difference in distance leads to the waves becoming out of phase, leading to alternate regions along the screen of constructive and destructive interferences.

What is more difficult to explain is why we get a diffraction pattern when light passes through only one slit. To explain this we need to bring in the Huygens' principle, that every point on a wavefront gives rise to a new wave. What this means is that as a wave passes through a single slit, there will be multiple new waves originating from points all across the slit. These new waves will interfere with one another, giving rise to a diffraction pattern even when there is only one slit.

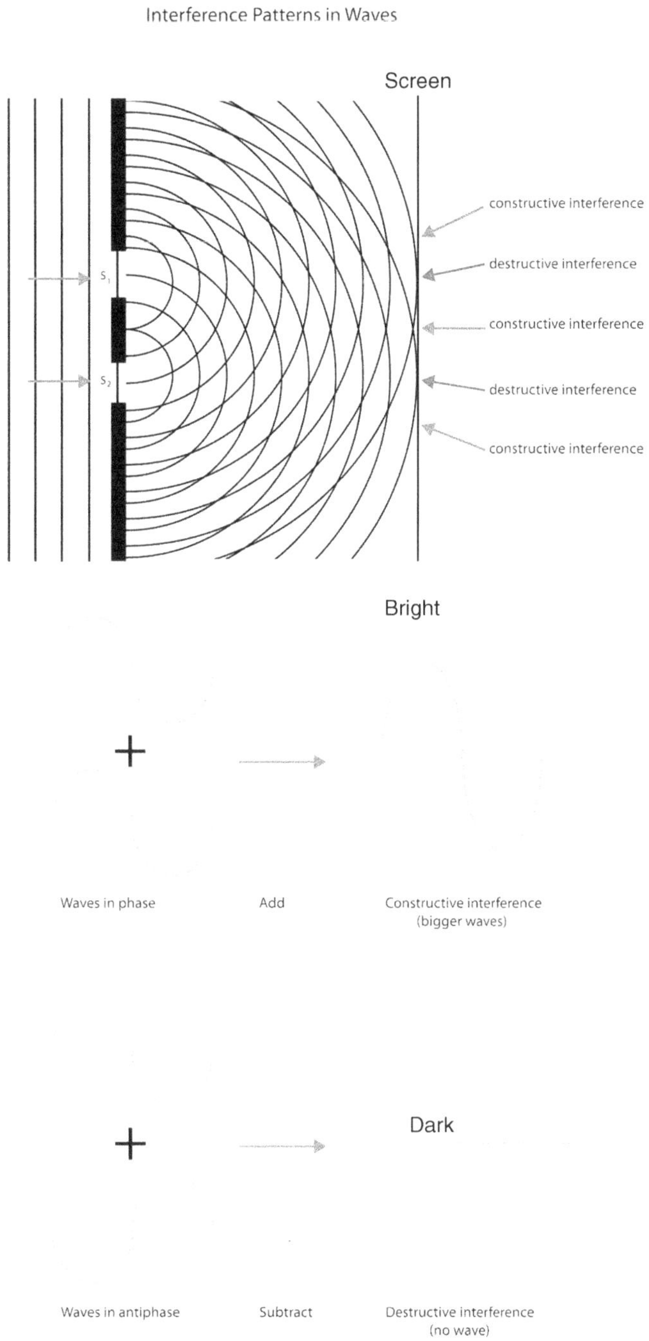

Steve Cymro/Shutterstock.com. Adapted by Kendall Hunt Publishing Company.

Figure 2.8. Diffraction and interference patterns. This figures shows examples of constructive and destructive interference. When two different waves are in phase (peak = peak, trough = trough) we get constructive interference. When waves are exactly 180 degrees out of phase (peak = trough) the waves cancel and there is no light. In the figure, an example is shown where two wavefronts from two different slits interfere. As waves travel from the slits, they travel different distances to the screen. At the point of the screen, waves traveling different distances become out of phase, so at some points along the screen we get constructive interference and at others we get destructive interference.

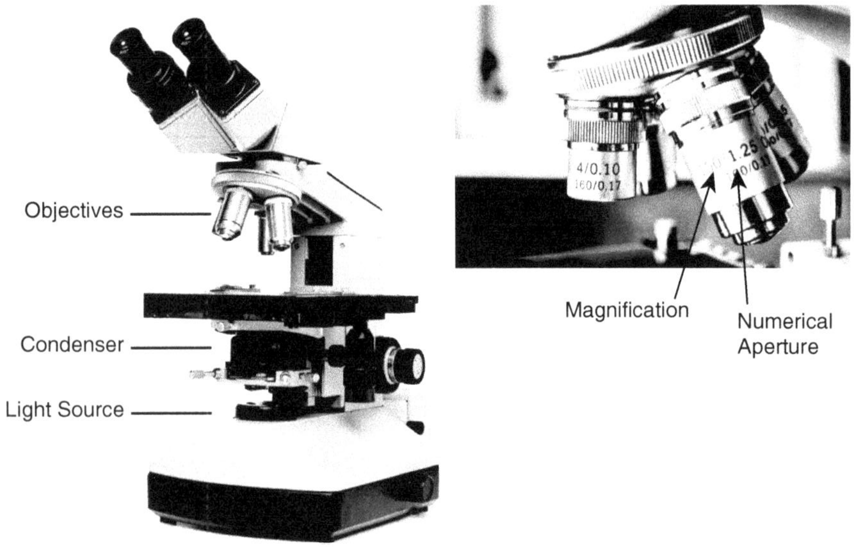

Figure 2.9. Imaging specimens with the light microscope. Left: A typical upright light microscope. Light is focused onto the specimen by the condenser and then captured by the objective lens. Right: A close-up of an objective lens shows its magnification (100X) and the numerical aperture (1.25), both being important features of the lens.

1) The Airy Disk

We have initially illustrated diffraction using a slit, but it is important to understand that any small object will cause light to bend around its edges and give rise to a diffraction pattern. If instead of slit we use a small pinhole, we get a circular diffraction pattern known as an Airy pattern. The Airy disk is seen as a central bright spot surrounded by dark and bright rings. Each object in the cell that interacts with light will generate points of light in the form of Airy disks.

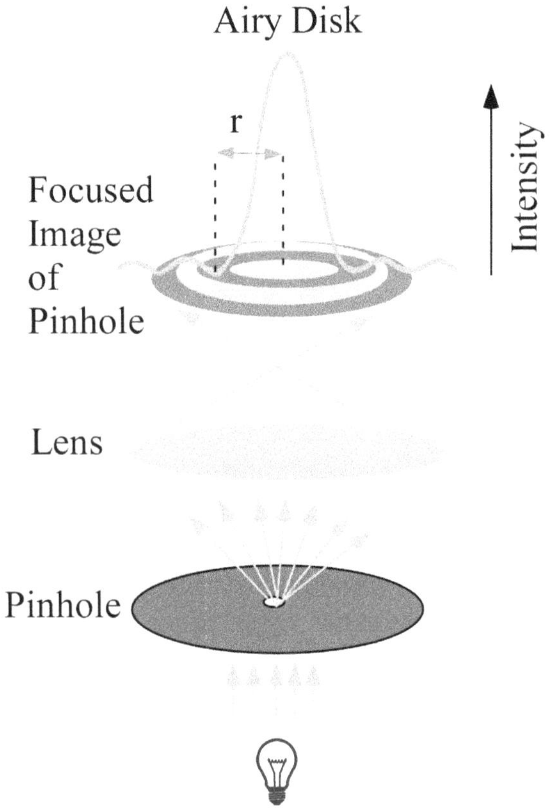

Figure 2.10. The Airy disk and Airy pattern. As light passes through the pinhole, it spreads out due to diffraction. When one tries to produce a focused image of the pinhole using a lens, the result is not a perfect image of the pinhole but rather a circular diffraction pattern. The central bright spot is known as the Airy disk. A series of light and dark rings surround the Airy disk. Together the central spot and associated rings are known as the Airy pattern. The Airy disk is the key to understanding the resolution we can obtain from a microscope. In particular we will be interested in "r," the distance between the central peak and the first minimum in the Airy pattern.

Source: M. Poenie. Lightbulb: okili77/Shutterstock.com.

In a science fiction movie, you might see a microscope that first observes a living cell and then zooms in to see molecules. This isn't really possible because of the properties of light, especially the phenomenon of diffraction. The size of the Airy disk and Airy pattern are fundamental to understanding how we are limited in what we can see using light microscopy. In the following sections, we will enumerate several principles that affect how the interaction of light with small cellular objects and the properties of microscope lenses limit what we can see.

2) Principle 1: The Size of the Airy disk diffraction pattern is related to the size of the object that generated it

The pinhole we are using here represents what happens when light interacts with small cellular objects. Basically the pinhole and small objects are similar in that they each generate Airy patterns from the diffracted rays. Here we extend this idea to say that for pinholes or objects that are near to or smaller than the wavelength of light, the smaller the pinhole, the wider the diffraction pattern of light coming off the object. As a result, the Airy disk diffraction pattern and surrounding rings get larger as the pinhole gets smaller. In order to detect the pinhole as a discrete spot, we must be able to capture and focus the rays coming out of the pinhole. If the rays spread out too far, the lens will not be able to focus the rays back into an image of the pinhole.

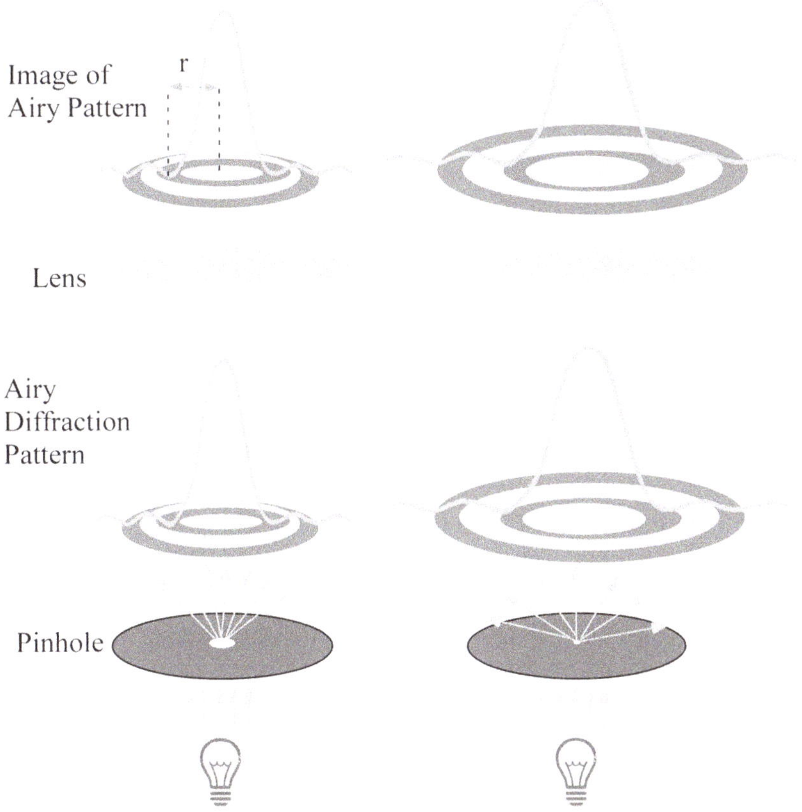

Figure 2.11. Principle 1. Pinhole size and the diffraction pattern. The smaller the size of the pinhole, the more light spreads out after passing through it. As a result the diffraction pattern of the pinhole becomes larger and blurrier as it gets smaller. The same is true of small objects in the cell. If the light spreads out at too large an angle, it cannot be captured and focused by the microscope objective.

Source: M. Poenie. Lightbulb: okili77/Shutterstock.com.

3) Principle 2: The Surrounding Rings of the Airy Pattern

The Airy pattern consists of the central bright spot (the Airy disk) and the surrounding rings or fringes of light. To fully understand the significance of these rings, we would need to discuss Fourier theory, which is outside our scope. But these fringes contain more detailed information about the pinhole image. To understand this we need to compare what an ideal image of the pinhole would look like and then see how the Airy pattern gives us an approximation of that ideal image (Figure 2.12). The fringe rings around the Airy disk correspond to finer details of the object and can be thought of as higher frequency information.

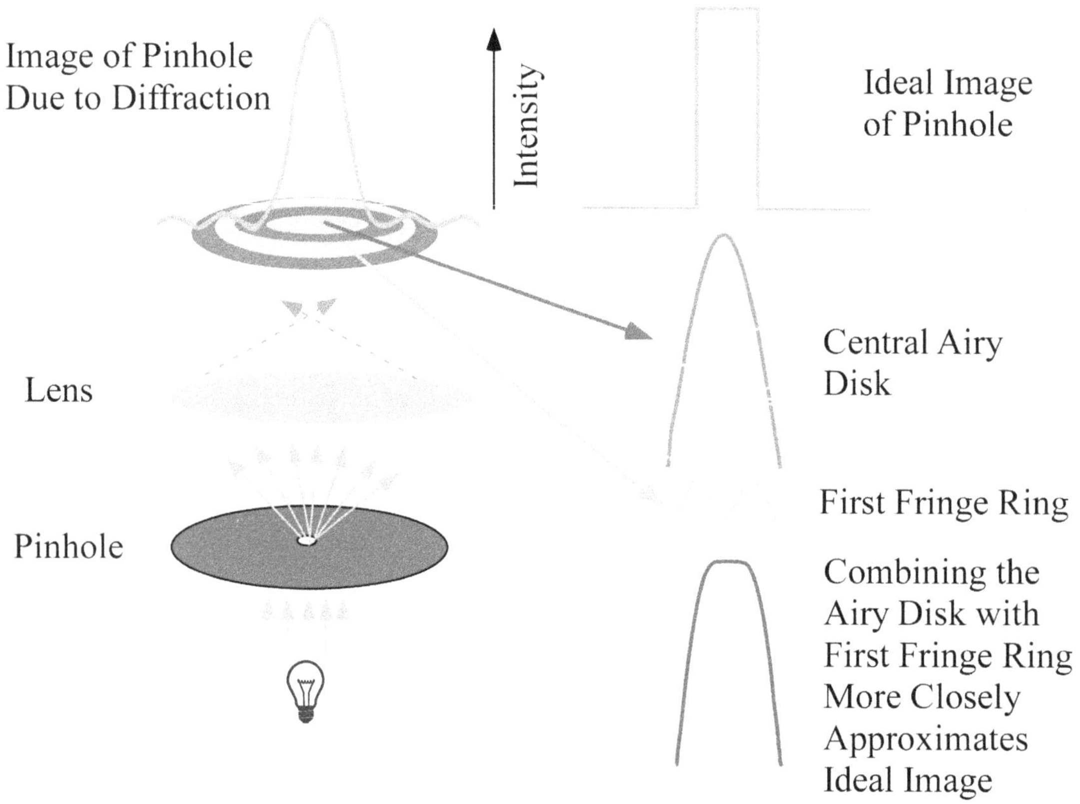

Figure 2.12. Principle 2. An ideal versus real image of a pinhole. For a pinhole filled with light, the ideal image would be a solid, uniformly bright central spot with totally dark edges. Instead we get the Airy pattern. The bright rings or fringes surrounding the Airy disk contain additional information about the pinhole image. To understand this we need to know that the real image is built by adding waves of different frequencies together. The central spot represents the large peak. The first fringe is a higher frequency wave that is superimposed on the large peak. When you mix these two waves together, you get a new wave that looks more square and gives a closer approximation of the ideal image. Each surrounding fringe adds high-frequency waves that add more detailed information, bringing us yet closer to the ideal image.

Source: M. Poenie. Lightbulb: okili77/Shutterstock.com.

4) Principle 3: A Light Microscope Image Is an Array of Airy Patterns

If each tiny object in a cell object generates its own Airy disk, then a microscope image actually represents a two dimensional array of Airy disks. Airy disks that are too close together will merge in the image and appear to be one object (Figure 2.13).

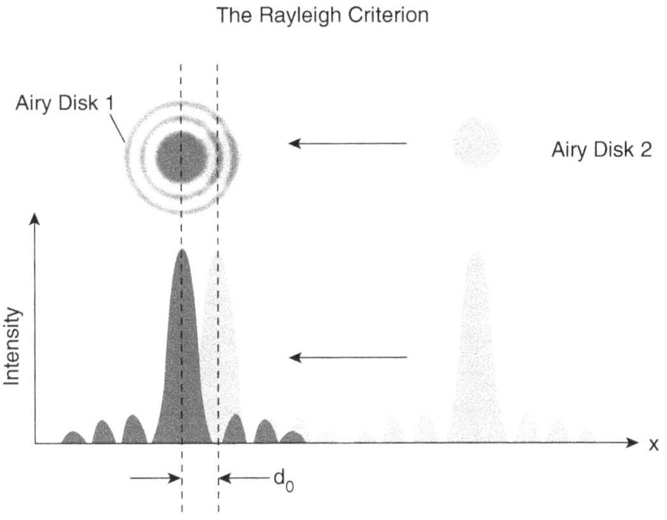

Figure 2.13. Principle 3. A microscope image is an array of Airy disks. An image of cellular components is actually seen as an array of Airy disks. Many of these Airy disks overlap. The ability to resolve small components of a cell boils down to being able to discern Airy disks as separate entities. When Airy disks overlap and merge together, the ability to resolve the objects that generated those Airy disks is lost.

5) Principle 4: The Diameter of the Airy Disk Determines the Limit of Resolution

The diffraction pattern that gives rise to the Airy disk determines the resolution of the image. If an objective lens could capture all the rays coming off the object we would have the best possible resolution. However, no lens can do this. Because the lens cannot capture all the rays that form the Airy Disk diffraction pattern the objective lens becomes the limiting factor in microscope resolution. We might say that the objective lens, due to its limitations can degrade the resolution available in the diffraction pattern. The size of the Airy disk determines the limit of resolution. The more narrow the Airy disk, the closer two disks can be together and still resolved as separate. The size of the Airy disk depends on the wavelength of light and by the numerical aperture (NA) of the microscope objective and condenser lenses. This is illustrated in Figure 2.14.

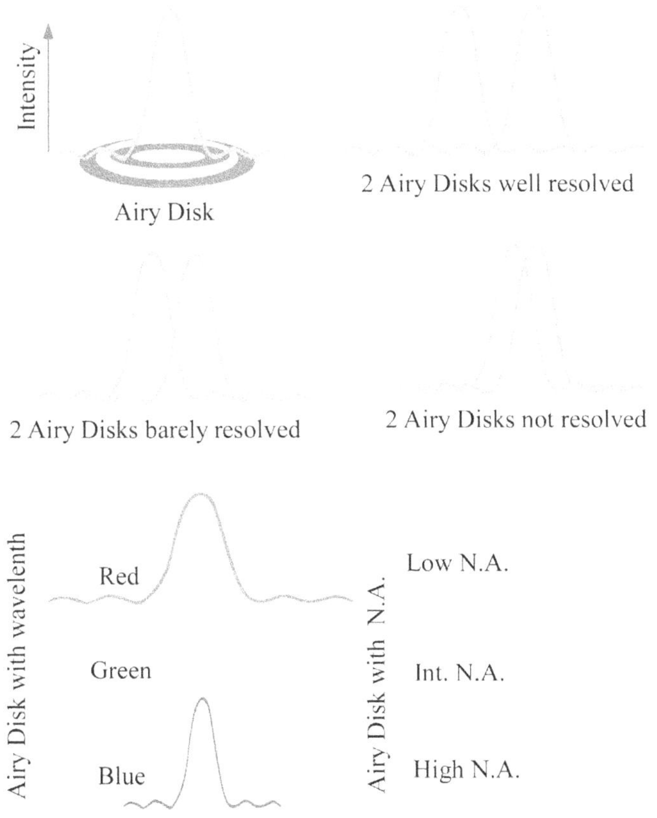

Figure 2.14. Principle 4. The size of the Airy disk determines the resolution of cellular objects. Resolution is defined as a distance: how close together two things can be and still be discerned as separate. In our case, we are referring to how close two Airy disks can be to each other before they appear to merge together and become one. In one definition of resolution, the intensity between the two peaks must drop by 50% in order to resolve the two Airy disks.

Source: M. Poenie.

6) Principle 5: The Wavelength of Light and the Numerical Aperture of the Microscope Lenses Affect Resolution through Their Affect on the Size of the Airy Disk

Resolution depends on the ability to discern two Airy disks as separate. This in turn depends on how small each Airy disk is. Other things being equal, the wavelength of light affects the size of an Airy disk. For a given size of pinhole or small object, the Airy disk will get smaller as the wavelength of light gets shorter. The numerical aperture of the lens (discussed in Section C.7) also influences the size of the Airy disk. Other things being equal, the higher the numerical aperture of the objective (and condenser lenses), the smaller the size of the Airy disk. As general principles, then, we can increase the resolution of the microscope by using shorter wavelength light and higher numerical aperture lenses.

7) Numerical Aperture

The numerical aperture of a lens refers to how wide a cone of light can be captured and focused by the lens. We noted in the previous section that the smaller the pinhole diameter (or object size), the wider the angle the rays of light will be spread out due to diffraction. In order to see an Airy disk as a discrete spot, we must, at a minimum, capture the rays contributing to that spot all the way out to the first minimum.

The numerical apertures of the microscope objective lens and the condenser are the key elements of the microscope that determine its resolution. We can begin the discussion of numerical aperture by presenting the formula for NA.

$NA = \eta \text{ sine } \theta$.

Here, theta (θ) is the largest half angle of diffracted rays that can be captured by the objective. The term eta (η) refers to the refractive index of the media between the coverslip and the objective lens Figure 2.16. This media could be air, water, or oil. The refractive index of the media is related to how wide a cone of light can be captured by the objective, because as light passes from one refractive index media to another, it can bend or, at wide enough angles (Figure 2.15), it will be reflected.

Figure 2.15. Refraction. As the pencil goes from one refractive index material (air) to another (water) it appears to be shifted. This is due to the phenomenon of refraction of light passing through media of different refractive indices.

Refraction limits the angle of rays that can be captured by the objective lens.

We can avoid the problem of refraction by using immersion oil between the glass coverslip and objective lens. Immersion oil has the same index of refraction as the glass lens and coverslip so the rays of light do not bend the way they do in Figure 2.15. Without immersion oil, some of the rays of light will not be captured by the objective lens so the half angle θ will be reduced compared to when immersion oil is present. The angle of rays that can be captured also depends on characteristics of the objective lens – how big it is, the kind of glass it is made from and other factors. These all contribute to the numerical aperture of the objective. This is important because the numerical aperture of the lens tells us not only how much light the lens can gather, but it also is related to resolution. The larger the NA, the narrower the Airy disk will be and therefore the better the resolution.

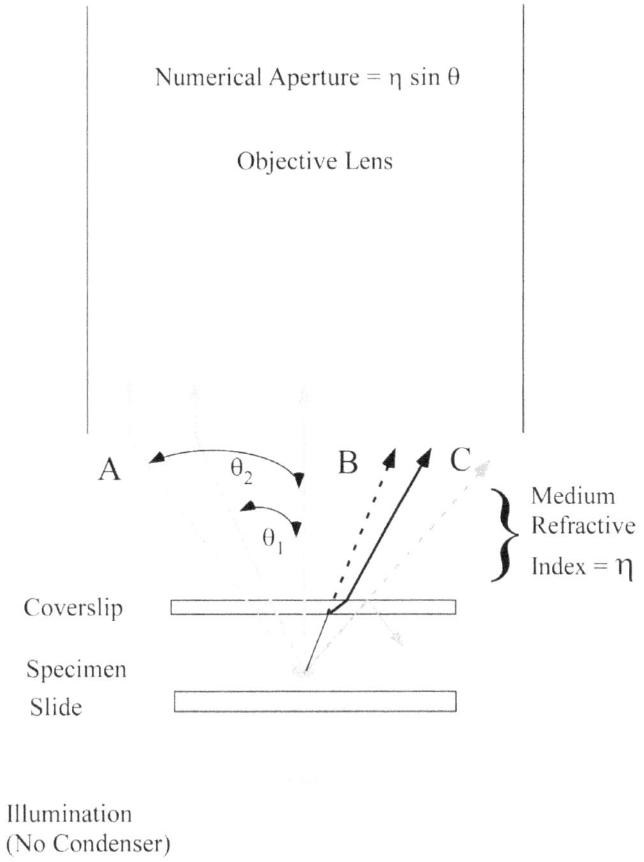

Figure 2.16. Lenses are characterized by their numerical aperture, which tells us how wide a cone of light the lens can capture. There are two components to the numerical aperture: theta (θ), the largest half angle of rays that can be captured by the lens, and *h*, the refractive index of the media between the coverslip and the lens. This media could be air, water, or oil. (A) Theta 2 represents a larger angle than theta 1, so a lens that could capture light coming off at an angle of theta 2 would have a larger NA than one that could capture only theta 1. (B) Due to refraction, light (solid line) passing from the glass coverslip (high refractive index) into air (low refractive index) would bend and might no longer be captured by the lens. Using oil, which matches the refractive index of glass, the rays would not be bent (dashed line). (C) At wide enough angles, light attempting to pass from the glass coverslip to air will be reflected instead of going straight. If there is oil as the medium instead of air, the index of refraction will be the same from the coverslip to the objective, and light rays will go straight to the objective (dashed line).

Source: M. Poenie.

8) The Effect of the Condenser Lens

A condenser increases the numerical aperture of the microscope lens system by focusing a cone of light on the specimen Figure 2.16. A condenser lens also has a numerical aperture. We can define a half angle for the cone of light coming from the condenser as α. Just as we did with the objective lens, we also need to take into consideration the refractive index of the media between the condenser lens and the glass slide. The numerical aperture of the objective lens and the condenser lens are additive, so that

NA of the microscope = NA objective + NA condenser.

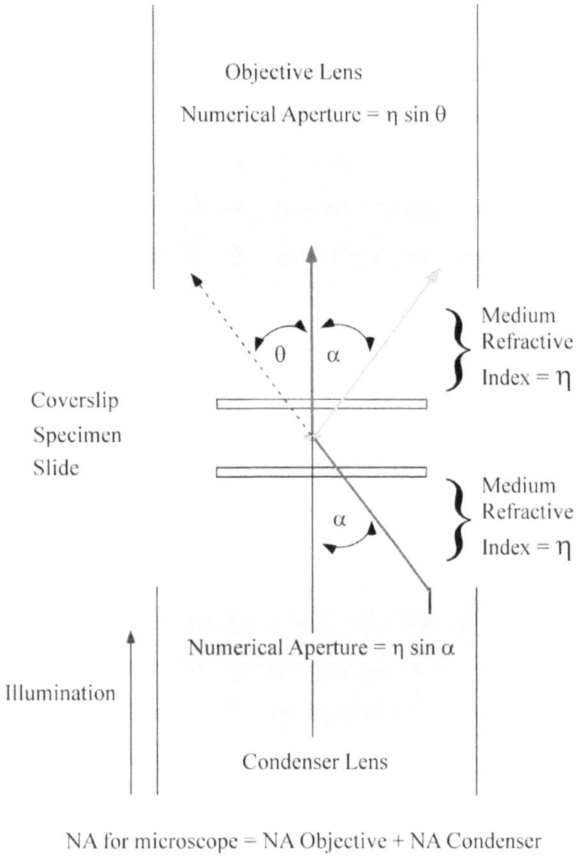

Figure 2.17. Effect of a condenser. With a condenser lens, instead of light coming perpendicular to the objective lens, it comes in as a cone focused on the specimen. The largest half angle of the cone of light is here shown as alpha (α). Without a condenser, the angle of light that could be captured by the lens was theta. Light coming from the condenser diffracted originally to angle theta is represented by the blue ray. In this case, the objective can now capture rays diffracted to larger angles than theta. The angles that can now be captured are the sum of theta + alpha and are represented by the orange line.

Source: M. Poenie.

D. The Formula for Resolution

We can now put the pieces together to get a formula that defines the resolution of the light microscope. Here r represents the distance between two small objects that can just barely be resolved:

Resolution without a condenser = r = 1.22 × λ / (NA objective) and

Resolution with condenser = r = 1.22 × λ / (NA objective + NA condenser).

There is a special case in fluorescence microscopy, which we will discuss in Section D.3. In fluorescence microscopy the objective lens functions as both the condenser and the objective. In this case, the formula becomes

r = 1.22 λ / (NA objective + NA objective).

Therefore,

r = 1.22 λ / 2 NA objective = 0.61 λ / NA objective.

In practice the limit of resolution for the light microscope is about 0.2 micron or 200 nm. In terms of magnification, this means that we can usefully magnify the cell by about 1500 times. While it would be easy to magnify the image more than that, any greater magnification would be called *empty magnification*. It would magnify a blurry image to a larger blurry image. You would not be able to obtain any more detail.

Current light microscopes have achieved their practical limits in terms of making the objective and condenser NA as large as possible. From the formula, the only thing we can do to improve resolution is to make the wavelength shorter. It is possible to operate a microscope using ultraviolet light but the optics have to all be made out of quartz, since glass is not transparent to UV. Ultraviolet light is also deadly to cells.

1) Contrast

One of the key things we need from a light microscope is resolution. But we need something else to make the microscope useful, and that is contrast. Cells are mostly water, and we might exaggerate a bit and say that looking at a living cell through the microscope is like looking at a drop of water in water. The point is that cells in water can be nearly invisible.

There are various ways to generate contrast. One can use light-absorbing dyes or stains to make certain structures of a cell stand out by absorbing light Figure 2.18. Stains are useful for certain large structures in the cell such as nuclei. It takes a lot of dye absorbed to cellular molecules in order make these molecules stand out. Staining works well to visualize nuclei, because there is a lot of DNA and RNA packed in a small space that will bind to certain types of dyes. For the most part, staining with colored dyes is not that sensitive a means for generating contrast.

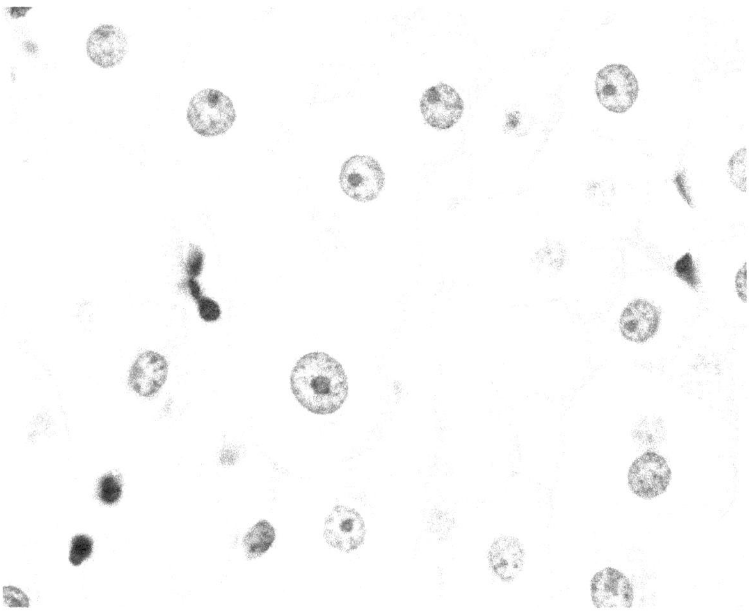

Figure 2.18 An image of liver cells (hepatocytes) stained with a dye that binds to DNA and RNA.

2) Optical Methods to Enhance Contrast

There are three main approaches to enhancing contrast of specimens under the light microscope. These are darkfield, phase contrast, and differential interference contrast, as illustrated in Figures 2.19, 2.20, and 2.21 respectively.

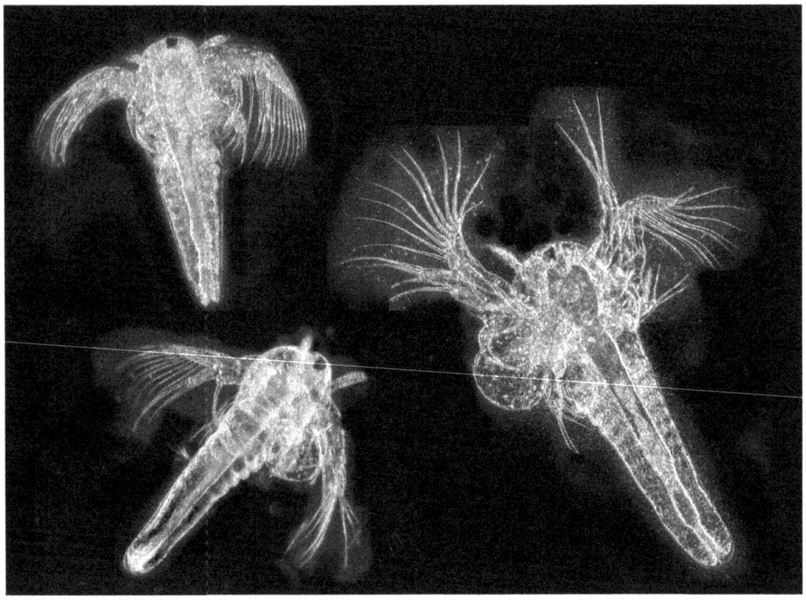

Figure 2.19. Darkfield microscopy focuses a hollow cone of light onto the specimen at angles that are too great to be captured by the objective. In the absence of a specimen, the field is completely dark. However, diffraction by the specimen leads to light leaving the cell at new angles that can be captured by the objective lens.

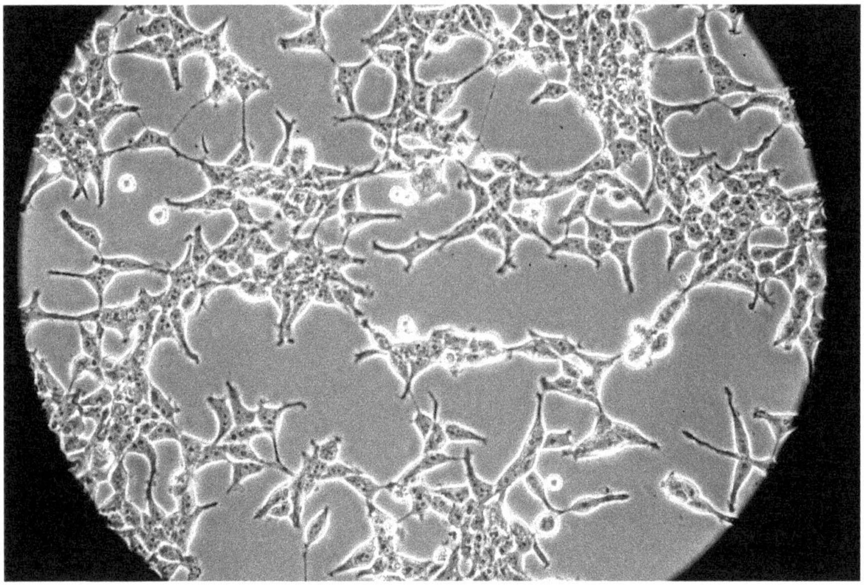

Figure 2.20. Phase contrast microscopy is based on the principle that the light waves are retarded to different degrees depending on how dense the material is that they go through. Light going through the specimen is separated into two different paths, one path that is undeviated and one that is bent due to diffraction. Light from these two paths is then recombined and they interfere with each other, generating bright or dark points depending on the relative phase of the two beams.

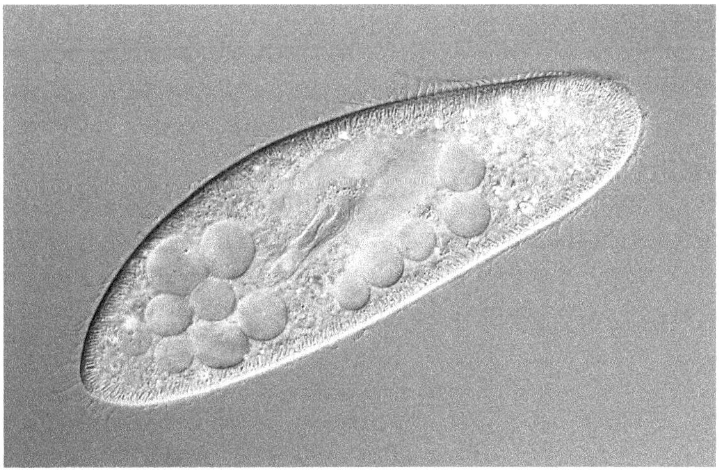

Figure 2.21. In differential interference contrast (DIC) microscopy, light is split into two spatially offset paths using a polarizer and a Wollaston prism. These two beams go through slightly offset positions in the cell. Variations in density can now alter the phase of one beam compared to the other. These two beams are then recombined with a second Wollaston prism. When recombined, the result is a difference in the polarization state of the light. This becomes a difference in brightness or darkness after the light passes through a second polarizer.

3) Fluorescence Microscopy

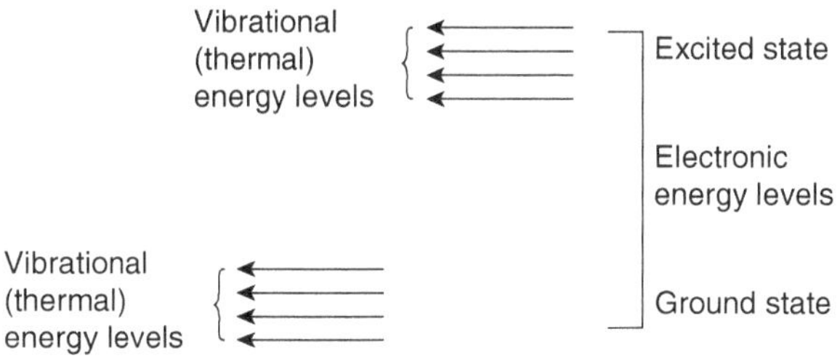

Figure 2.22.

Molecules exist in different energy states. These can be states of motion (vibrational energy levels) related to heat. Molecules can jump to much higher energy levels due to their electrons jumping from one orbital to another (Figure 2.22). We say that the energy levels a molecule can occupy are quantized because molecules can only jump (or fall) from one energy level to another. There is no smooth or gradual change from one energy level to the next. In the lowest energy state (the ground state), molecules will exhibit different types of motions. We might say that these motions correspond to how "hot" the molecule is. However, when molecules absorb photons whose wavelengths correspond to the visible light, this causes electrons to jump from an inner orbital to one farther out. This is called an electronic transition and involves much larger changes in energy than moving from one vibrational level to another.

E. Absorption and Fluorescence

Photons can be absorbed by molecules, causing the electrons of a molecule to jump to a higher energy level. For fluorescent molecules, also known as fluorphores, this is known as excitation. Likewise, the electrons can fall back to a lower energy level and release a photon which is called emission. As with absorption of a photon, when a molecule falls back to a lower energy level, it must fall from one of the allowed steps to another lower step Figure 2.23.

The basic principle of fluorescence is the absorption and reemission of light. However, this idea needs refinement. The transition to an excited electronic state can also cause the molecular bonds within the molecule to vibrate. Here we might think in terms of the molecule becoming hot when it undergoes an electronic transition. With time, some of this heat will be transferred to the surroundings and the molecule will cool off. This corresponds to the molecule going from higher to lower vibrational energy levels while it is still in the excited state. At some point (usually within nanoseconds) the electron will fall back to a lower orbital such that the molecule returns to its ground state. This transition once again can result in molecular vibrations such that when a molecule falls back to the ground state it ends being hot. The key point is that due to the loss of some energy as heat, the energy of the absorbed photon is greater than that of the emitted photon (Figure 2.24). Since the energy of a photon corresponds to its wavelength or color, this means that

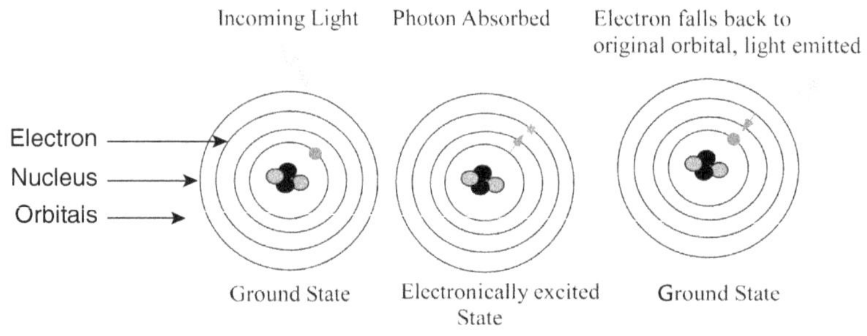

Figure 2.23. Absorption and reemission of light. When a photon of light in the visible wavelength range is absorbed, it causes an electron to jump from a lower energy orbital to a higher energy orbital. While the electron is in a new orbital, an atom is said to be excited or in an electronically excited state. The electron eventually falls back to its normal orbital. When this happens, the energy difference between the exited state and the ground state is released as a photon of light.

Source: M. Poenie.

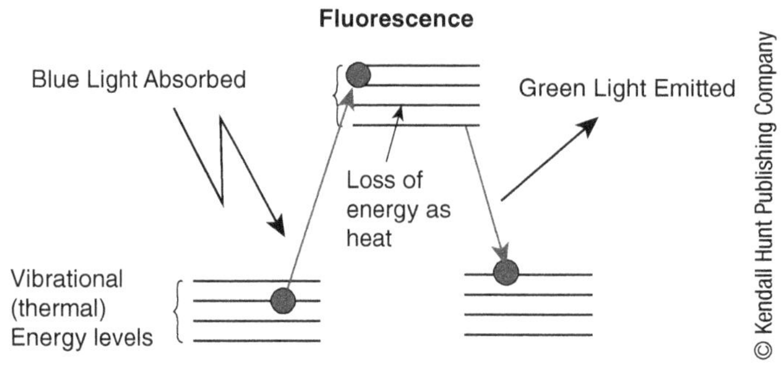

Figure 2.24. Excitation and Emission Molecules can exist at different energy levels. The lowest energy is the ground state. However, molecules at the ground state can have different energy levels due to motions of atoms within the molecule or the molecule as a whole. The energy needed to cause these motions corresponds to heat so we might call these thermal energy levels. When a molecule absorbs a photon in the visible wavelength range, it can cause electrons to jump from one orbital to another which represents a much larger change in energy. This is known as an electronic transition where the molecule jumps in energy from the ground state to an electronically excited state. Molecules usually do not spend much time in an electronically excited state because the electron falls back to its original orbital and in doing so emits a photon of light. When a photon is absorbed, and an electronic transition takes place, a change in vibrational levels is also involved such that the molecule is hot. This heat is released to the environment before the electron falls back to its original orbital. After the electron returns, once again the molecule becomes hot. The result is that the transition to the excited state represents a bigger change in energy than the return to ground state, the difference being the heat energy.

the light absorbed has a shorter wavelength than the light emitted. The difference between the wavelength of the absorbed light and the emitted light is known as the Stokes shift (Figure 2.25).

1) The Design of the Fluorescence Microscope

Fluorescence provides the basis for one of the most sensitive means of detecting molecules in the cell. It is possible to detect a single fluorescent molecule by its fluorescence. However, to take full advantage of the power of fluorescence, we need to utilize the special optical arrangement of a fluorescence microscope. The key to the sensitivity of fluorescence microscopy is that (ideally) we see the fluorescence of molecules in the absence of background light. We can illustrate this sensitivity using stars in the night sky. During the day we cannot see them due to the background of daylight. Yet at night we see them clearly.

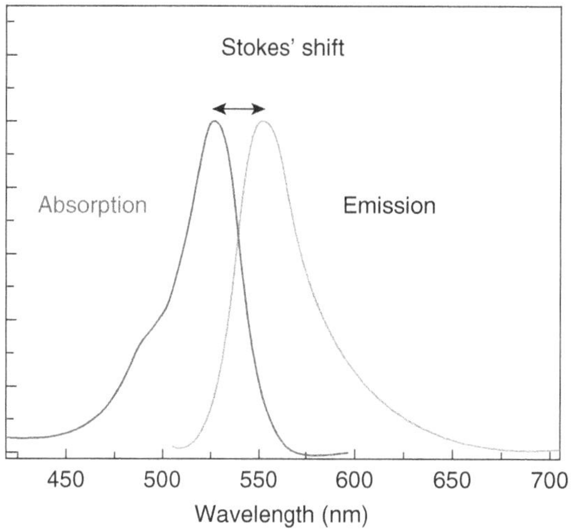

Figure 2.25. The Stokes shift. Fluorescent molecules absorb light best at particular wavelengths. Using instruments such as a spectrophotometer or spectrofluorometer one can scan through a range of wavelengths and determine how much light is absorbed at each wavelength. The result is an absorption (or excitation) spectrum. There are also preferred wavelengths where a fluorescent molecule that, once excited, will emit light. By exciting a fluorophore at a constant excitation wavelength one can scan the wavelengths the emitted light to determine to determine the wavelengths where maximal emission occurs. Thus a fluorophore will exhibit a peak excitation wavelength and a peak emission wavelength. The difference between these two peaks (in nm) is known as the Stokes Shift.

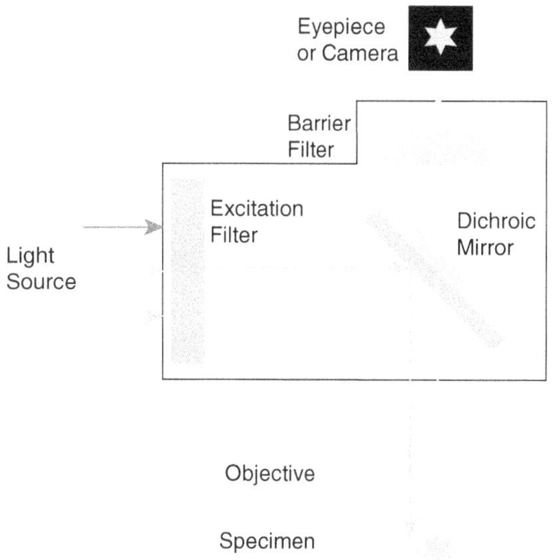

Figure 2.26. The fluorescence microscope. The fluorescence microscope is designed to excite the fluorescent molecules in the specimen with bright light corresponding to its peak excitation wavelength but eliminate the excitation light from the field so that one sees only the fluorescence. This is done by starting with a very bright white light source and using a filter that allows only the excitation wavelength (here it is blue) to be transmitted. This excitation light is then reflected down to the specimen by a dichroic mirror. Dichroic mirrors are optical elements that reflect at one wavelength and transmit at another. The excitation light then excites fluorescent molecules in the specimen such that it glows green. The green light is now collected by the objective lens and passes through the dichroic and then the barrier filter that allows only green light to pass through. The result is that unless there are molecules fluorescing at the right wavelength, the field will be dark because any blue excitation light has been blocked.

2) The Confocal Fluorescence Microscope

Fluorescence microscopy is an important too but it suffers a serious limitation. When the excitation light is focused onto the specimen, it will be brightest at the focus but will travel through the specimen as a first converging and then diverging cone of light. Light in any part of the converging and diverging cones can excite fluorescence in the specimen. However, the fluorescence from these regions will either be above or below the plane of focus. Therefore, they will not be focused into an image but instead they will become part of a blurry background. The special optical configuration of the confocal fluorescence microscope is designed to eliminate out-of-focus light from the image (Figure 2.27).

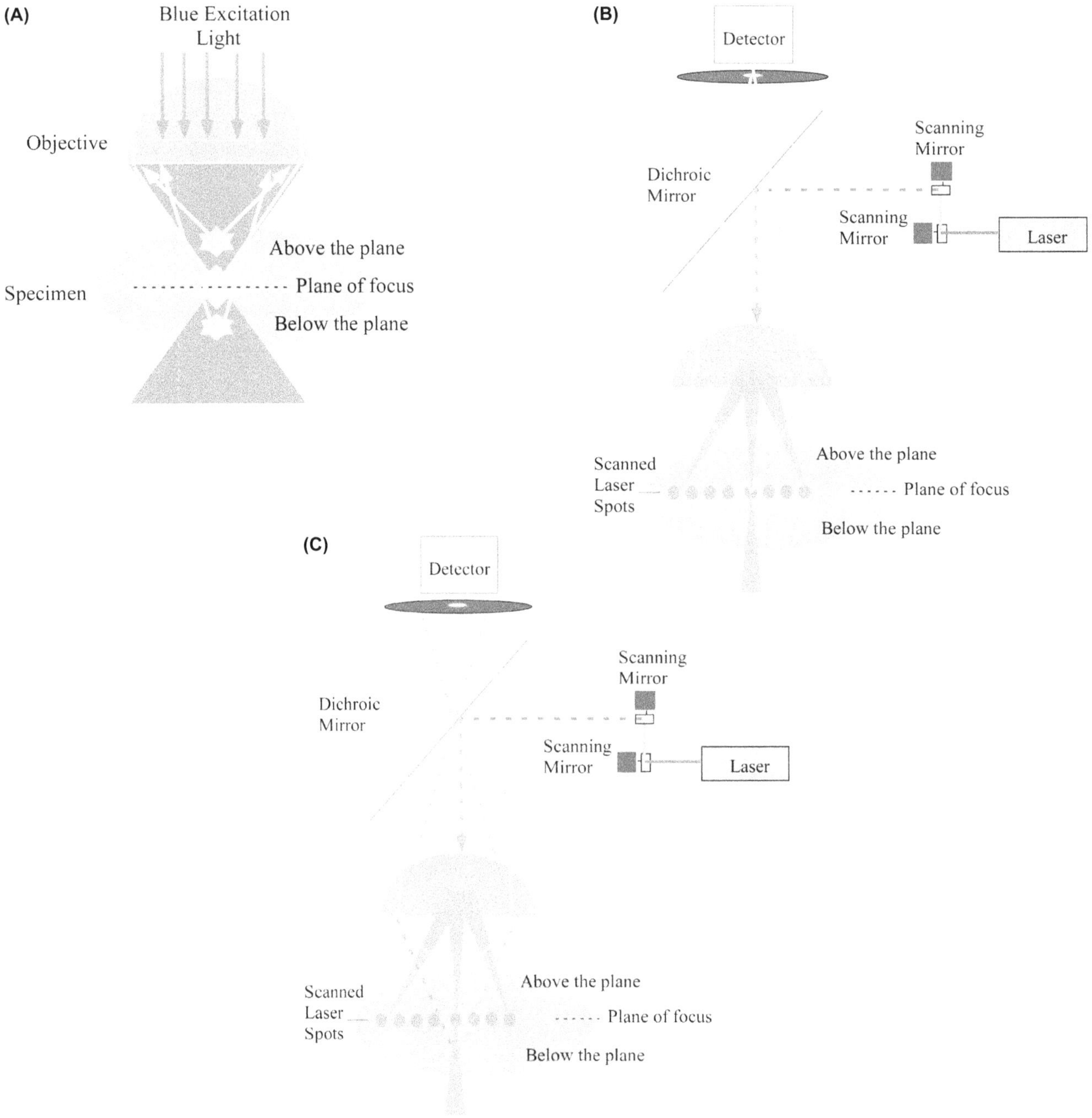

(continued)

◀ **Figure 2.27.** Wide field versus confocal microscope. (**A**) In a standard (wide field) fluorescence microscope, the excitation light (blue) is focused through the specimen as a converging and then diverging cone of light. Fluorescent molecules anywhere within the illuminated area can be excited and emit (green stars), although the strongest excitation is at the focus. All the emitted light can be captured and become part of the image, but light that comes from regions above or below the focus will contribute to a blurry background. (**B**) In a confocal fluorescence microscope a laser beam is focused through a pinhole and then onto the specimen. This results in the illumination of a small spot on the specimen rather than illuminating the whole specimen. This illuminated spot will be scanned back and forth across the specimen to produce an image. (**C**) While illumination with the laser illuminates only a spot on the specimen, the illumination still excites fluorescent molecules above and below the plane of focus. However, fluorescence originating from above or below the plane of focus is largely blocked from passing back through the pinhole. Fluorescence originating from above the plane of focus would come to a focus at a point beyond the pinhole. Fluorescence originating from a point below the plane of focus will converge to a point before the pinhole. In both cases, most of the out-of-focus light is blocked.

Source: M. Poenie

F. We Can See Things That We Cannot Resolve

We can use stars again to illustrate that we can see things we cannot resolve. The biggest telescopes in the world cannot resolve stars – but you can see them. However, the image of a star will be much bigger than it should for a luminous object that far away. It is similar viewing points of light under the microscope. In Figure 2.28, the size of a single fluorescent protein is shown (central green spot) relative to the size of the focused Airy disk that is seen through the microscope. The key concept is that no matter how small the source of light is, it can never be focused to a spot smaller than the limit of resolution. The actual size of the spot (Airy disk) will depend on the wavelength of light and the NA of the microscope objective lens.

Often, researchers want to visualize where particular proteins are in a cell. Protein diameters vary over a range, but for purposes of illustration we will say that an average protein is 5 nm in diameter. The limit of resolution is about 200 nm so a protein is much too small to be resolved. Nevertheless, it possible to see the fluorescence of a single fluorescent protein molecule, but the Airy disc obtained will be no smaller than the limit of resolution.

G. The Electron Microscope

The electron microscope provides a way to image cells and cell components at a much higher resolution than with light microscopy (Figure 2.29). Instead of using light, imaging is done with an electron beam, as the name of the microscope suggests. An electron beam differs from a beam of light in that electrons will collide and be scattered by air, and so the beam can be transmitted only in a high vacuum. However, as with light, electrons have a wavelength that varies with their energy. The energy of electrons can be increased by accelerating them through a high-voltage field. In a transmission electron microscope, this can range from around 100,000 volts up to 1 million volts. As the electrons are accelerated by the high voltage, their wavelength gets shorter – much shorter than the wavelengths of visible light.

At 200,000 volts, the wavelength of the electron is 2.5×10^{-12} meter, which is about 200,000 times shorter than that of green light. The formula for resolution used for light microscopes still applies, so with a much shorter wavelength, one should obtain much better resolution. This is partly offset

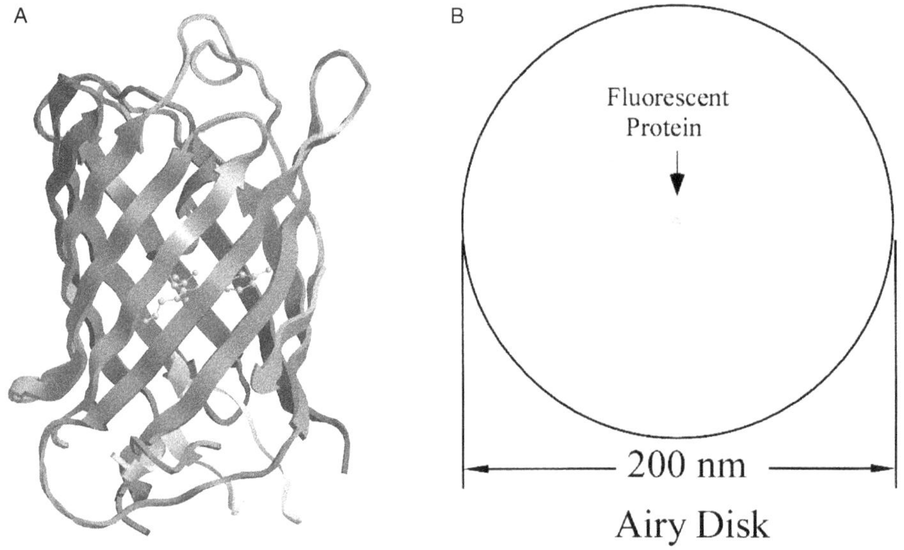

Figure 2.28. (**A**) Green fluorescent protein (GFP) is made by the jellyfish *Aequoria victoria*, causing it to glow at night. This and other fluorescent proteins have become popular tools to track proteins in living cells. Using molecular biology techniques, it is possible to fuse together a protein of interest with GFP. This is usually done by obtaining DNA copies (cDNA) of the mRNA for each protein. These cDNAs can be spliced together and then reintroduced into the cell where they will be transcribed and translated as one protein. The product of fusing two different genes together is known as a chimera When the chimeric protein is expressed, its fluorescence can be seen in the microscope. (**B**) We can see a single GFP molecule in the fluorescent microscope but the image (Airy disk) that we see is much larger than the protein itself. Due to the limit of resolution, we cannot focus a point of light to a smaller spot than that dictated by the resolution of the microscope optics.

A: Raimundo79/Shutterstock.com; B: Source: M. Poenie.

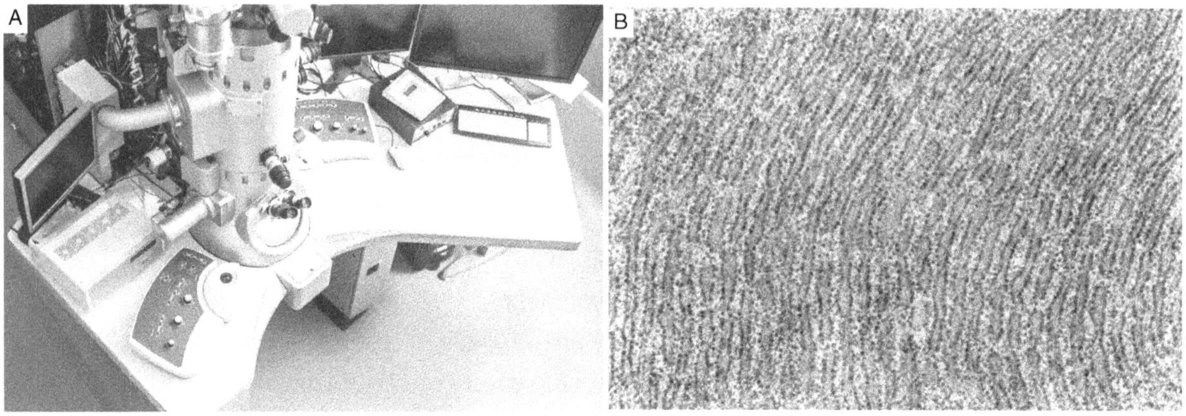

Figure 2.29 Electron microscope and an electron micrograph. (**A**) A standard electron microscope. (**B**) An electron micrograph of the endoplasmic reticulum (ER). Along the membranes of the ER there are many tiny black dots in the image. These are mostly ribosomes, which are far too small to resolve in the light microscopy.

A: Elizaveta Galitckaia/Shutterstock.com. B: Jose Luis Calvo/Shutterstock.com by the poor performance of magnetic lenses used to focus the electron beam. These lenses have relatively small numerical apertures and tend to have aberrations as well. Despite these problems, electron microscopes have achieved resolutions down to 0.4 angstroms, which is adequate to see individual atoms.

H. Superresolution: Overcoming the Limit of Resolution

Despite our discussion of the limits of resolution using the light microscope, there are actually various optical methods that can be used to overcome the limit. development of these methods, The development of these methods led to Nobel Prizes for the inventors Eric Betz, Stefan Hell, and William E. Moerner.These methods are described below.

1) PALM/STORM

PALM stands for "photoactivated localization microscopy." STORM is another name for essentially the same concept, where one breaks the limit of resolution by switching the fluorescence of fluorescent molecules on and off. The idea is to record an image of a field that contains a relatively small number of fluorescent molecules such none of them overlap. The white blob corresponding to each fluorescent molecule is much larger than the size of the fluorescent molecule itself. However, from the image, one can accurately determine the center of the fluorescent blob within nanometers. Suppose then that within computer memory space we can create an image with nanometer resolution. Proteins are less than 10 nanometers in diameter. So with the accurate knowledge of the center of each blob and knowing that it comes from a molecule a few nanometers in diameter, we can now position bright dots of nanometer dimension in our virtual image.

The key to this technique is that we can switch off the fluorophores using one wavelength of light and then switch on a new set of fluorescent molecules. The specimen may contain tens of thousands of potentially fluorescent molecules. With each cycle we switch off the previously recorded fluorophores and then we switch on a new set in the image Figure 2.30. With each cycle, the centers of the individual fluorescent spots are determined and their positions are added to our virtual image. By doing this over and over, we can build up a picture that is based on where all the fluorophores are with each protein represented by its real size.

2) Stimulated Emission Depletion (STED)

STED is an entirely different approach to obtaining superresolution. It is a laser scanning technique that uses two lasers to generate an illuminated laser spot that is much smaller than obtained from the single laser normally used in scanning microscopy. It begins with a conventional laser scanning approach to excite fluorescence in the specimen. Assume in this case that we are exciting fluorescence with a green laser. The focused spot will be of typical size, diffraction limited or larger. What is novel about STED is that it uses a second laser to superimpose a doughnut-shaped spot on top of the first laser (Figure 2.31). While the purpose of the first green laser is to excite fluorescence, the purpose of the superimposed doughnut-shaped beam is to cause excited molecules to lose their energy as heat instead of giving off light. The only region that is left excited is the central region of the doughnut, which is not affected. The diameter of the central region of the doughnut is much smaller than the limit of resolution. As a result one obtains fluorescence from a scanning spot that is only approximately 50 nm in diameter.

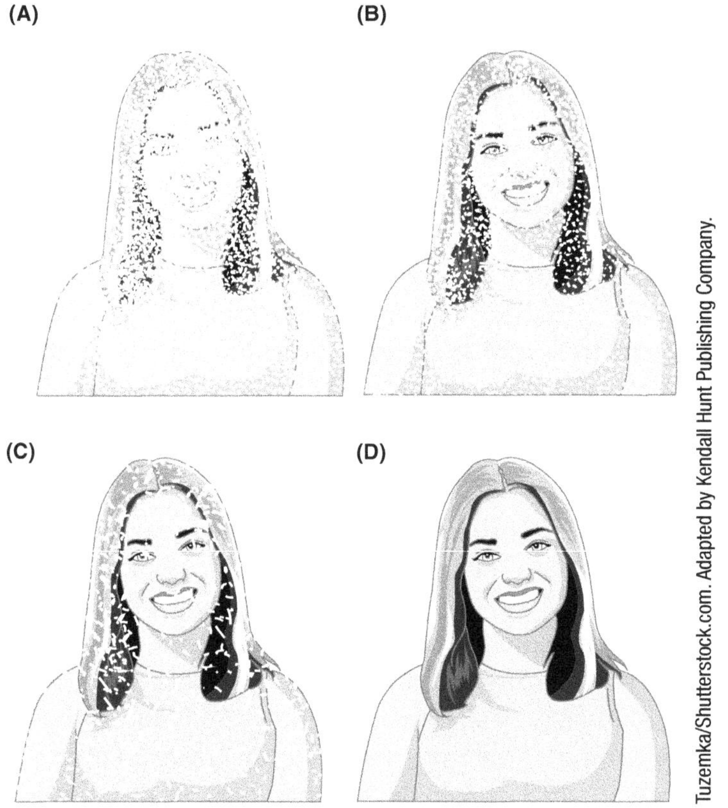

Figure 2.30 PALM. By repeated photoactivation/photoinactivation, one builds up an image from the computed centers of individual fluorescent proteins.

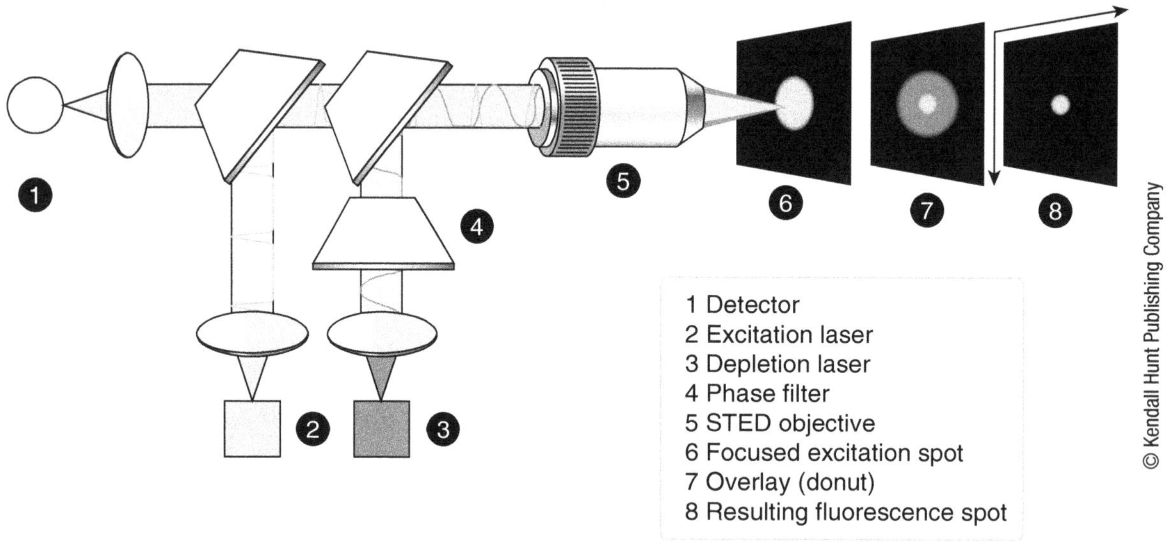

Figure 2.31 Fluorescence is excited to produce a typical illuminated spot, which one follows with a pulse of longer wavelength light formed as a doughnut around the initial illuminated spot. This stimulates emission, leaving only the small central region of the doughnut fluorescent.

2 Immunofluorescence: Combining Antibodies with Fluorescence

Immunofluorescence is one of the most important tools of cell biology. It allows us to identify the cellular location of specific molecules, including proteins, carbohydrates, and even some lipids. To accomplish this, immunofluorescence brings together the sensitivity of fluorescence as a method of detection and the ability of antibodies to distinguish one kind of protein from the estimated 100,000 different proteins in the cell.

A. Antibodies

Antibody molecules are made by vertebrate animals in response to foreign molecules in the body. They are produced by B lymphocytes, which represent one arm of the immune system, the other major arm being the T cell system. Although there are different classes of antibodies, for our purposes we will consider IgG (immunoglobulin G) as our typical example. Antibodies are built from four peptide chains that are joined through disulfide bonds. Due to differences in length of the chains, two of the peptide chains are referred to as heavy chains and the other two are called light chains (Figure 2.32). At one end of each heavy and light chain is a constant region, while the other end is a variable region. The constant end of the paired heavy chains is known as the Fc end of the antibody molecule.

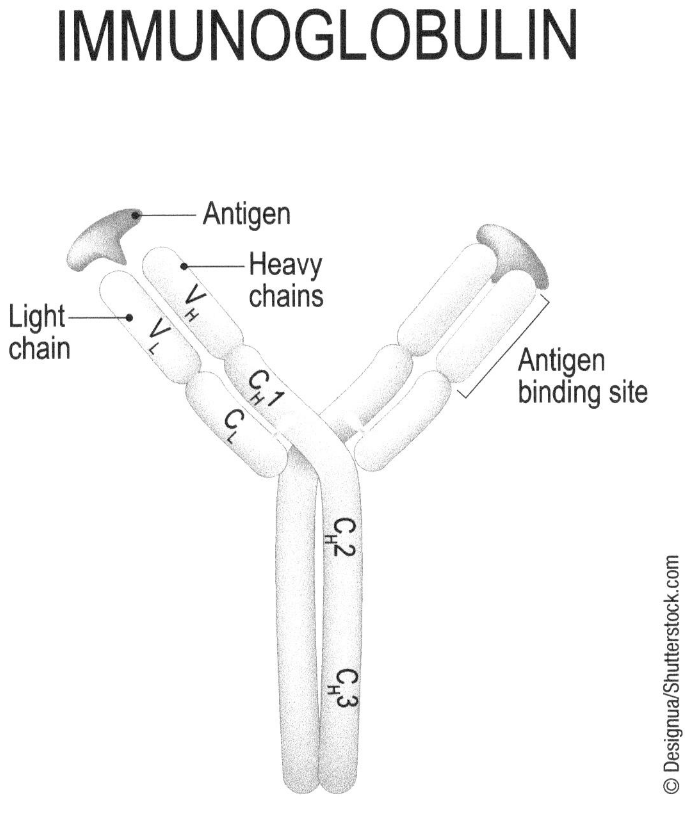

Figure 2.32 An antibody molecule. This antibody molecule is composed of two heavy chains and two light chains. Each of these chains contains a constant end (C) and a variable end (V). The constant end of the paired heavy chains is known as the Fc end, whereas the paired variable chains are the Fab end of the antibody.

Antibodies work by binding to foreign molecules which are called antigens, and the binding site for the antigen is created by the combined variable regions of a heavy and light chain. The antigen binding site is referred to as the Fab portion of the antibody. Each whole antibody molecule has two antigen binding sites (i.e., it is bivalent), which will be important in future discussions.

B. Antigens

Antigens are substances foreign to the body (proteins, carbohydrates, lipids, etc.) that trigger an immune response or reaction. Antibody genes are formed by a combinatorial process that links gene segments known as V (variable), D (diversity), and J (joining) together to form a functional antibody gene. B cells then display this antibody on their surface, where it can be exposed to antigens. Importantly, odds are that each naïve B cell will make a unique antibody that is different from that made by other B cells. When the population of B cells is exposed to an antigen, one or more B cells will bind to the antigen. The binding of the antigen triggers a B cell to multiply and mature either into a memory cell or a plasma cell. It essentially clones itself, generating many copies of the original cell. Plasma cells will then secrete antibody into the blood where it can encounter and bind to the antigen.

In the case of protein antigens, it is possible or even likely that more than one B cell will display an antibody that binds to the antigen, but most likely they would not bind to the same site on the protein. Antibodies on average bind to a segment of a protein that is only 6 amino acids long, and this binding site is known as the epitope. Proteins, however, can be hundreds of amino acids long, so along the length of a protein there are many possible epitopes.

1) Polyclonal Serum versus Monoclonal Antibodies

As mentioned above, an antigen may stimulate more than one B cell to activate, clone itself, and secrete antibodies into the blood. As a result, the blood can contain a mixture of different antibodies that bind to different epitopes on the same antigen. Early on, the only way to obtain antibody against a particular antigen was to immunize an animal with repeated injections of the antigen. During the period of immunization, antibody against the antigen would build up in the blood, then blood would be drawn and allowed to clot. After a while the clot retracts, oozing out an amber colored fluid called serum. The serum represents the soluble fraction of the blood devoid of cells and platelets. This serum is called a polyclonal serum because it contains antibody from more than one type of B cell.

During the 1970s, Cesar Milstein and colleagues developed a new approach to making antibodies, which led to the 1984 Nobel Prize in medicine. Their goal was to isolate individual B cells that could be grown into a large number of cells and that would produce a particular antibody. The trouble was, B cells obtained directly from an animal do not grow and multiply. However, Milstein et al. discovered that if they fused a B cell with a myeloma cell (a B cell tumor), the fused hybrid would multiply forever and secrete antibody into the medium. To prepare antibodies they would then immunize a mouse, isolate B cells, and fuse them with myeloma cells. These cells were treated with drugs that allowed only hybrids to survive, and then they were diluted out into multiwell plates so that on average, there would be only one hybrid per well. After a while, the cells would multiply and secrete enough antibody into the wells to be tested. The solution in the wells would then be screened for antibodies that bound to the antigen. From the positive wells one could make essentially endless amounts of antibody. This antibody is called a monoclonal antibody because it is derived from only one B cell clone.

C. Immunofluorescent Staining

We can use antibodies to visualize where proteins are in a cell. One can obtain an antibody against a particular antigen and chemically attached a fluorescent molecule to it, so that anyplace the antibody binds will show up as a fluorescent region. This is known as direct immunofluorescence. More commonly, two antibodies are used: first the primary antibody that recognizes the antigen, and second a secondary antibody that is fluorescently labeled and which binds to the primary antibody. This is known as indirect immunofluorescence.

Antibodies are normally made only against foreign molecules. Therefore, if we want to make antibodies against a mouse protein we must make those antibodies in another animal. A mouse normally cannot make antibody against mouse proteins. It is important, in referring to antibodies, to know where the antigen came from (what animal species) and what species of animal was used to make the antibody. For example, if tubulin from humans was used to immunize a mouse, the resulting antibody would be called mouse anti-tubulin antibody. This would be the primary antibody used to localize tubulin in cells.

In indirect immunofluorescence (Figure 2.33) the fluorescently labeled secondary antibody binds to the primary antibody. Since in our illustration, the anti-tubulin antibody was made in mice, the secondary antibody must bind to mouse antibody. For this, the secondary antibody would have to be made in another animal such as a goat, horse, or rabbit. If it were made in a goat, it would then be called goat anti-mouse Ig (Ig refers to immunoglobulin or antibody).

There are two main reasons why indirect immunofluorescence is preferred for most applications. The first is a matter of convenience and economy of scale. Primary antibodies are typically expensive and purchased (or prepared) in tiny amounts. It is difficult to fluorescently label and purify such tiny amounts of antibody. In starting with small amounts of protein, one might not have any left after trying to purify it away from the unreacted fluorescent label. A mistake can be very expensive. On the other hand, companies can economically make large amounts of fluorescently labeled secondary antibody that many people can buy.

The second advantage of indirect immunofluorescence is that it is more sensitive than direct immunofluorescence. We mentioned that one protein can have many epitopes. If we immunized a goat with mouse antibody, the polyclonal goat serum could contain many different antibodies that bind to the mouse antibody. Having multiple fluorescently labeled secondary antibodies bind to the primary antibody multiplies the number of fluorescent molecules connected to the antigen, thus giving a brighter signal.

The procedure for immunofluorescence is relatively straightforward. One first must "fix" the cells. Fixation involves stabilizing proteins in place in the cell so they do not wash away. Fixatives can be solvents like methanol or formaldehyde. After fixation the cell must be permeabilized. Antibodies are large molecules that cannot go through the intact cell membrane. Permeabilization involves using solvents or detergents that open holes in the membrane. After permeabilization the cells are incubated with primary antibody, which binds tightly to the antigen. Any unbound antibody is then washed away. One then adds the secondary antibody, which binds tightly to the primary antibody, and the cells are washed again. Finally, cells are mounted for viewing in the microscope. An example of a cell labeled with antitubulin using the indirect immunofluorescence technique is shown in Figure 2.34.

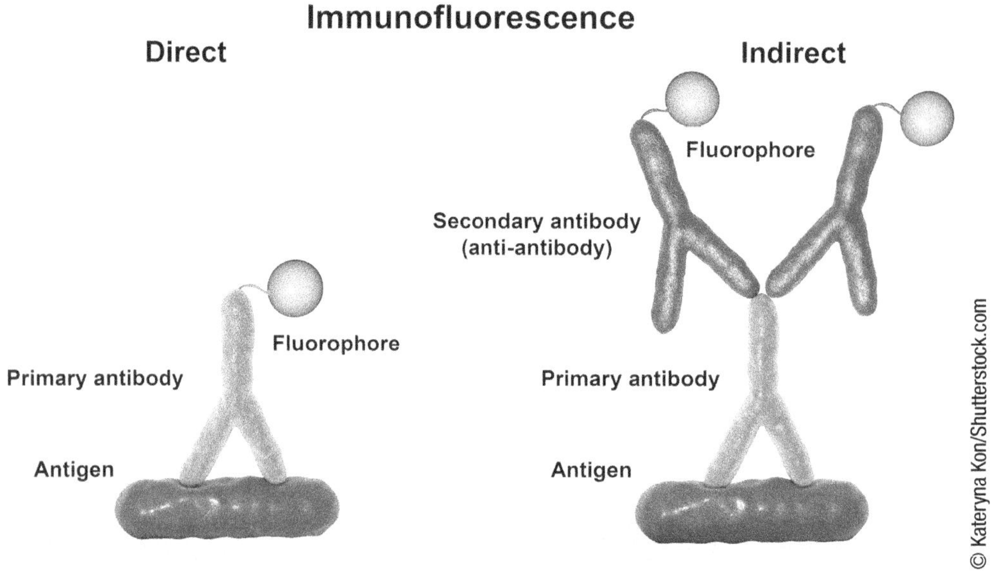

Figure 2.33 Immunofluorescence can be direct or indirect. In direct immunofluorescence the primary antibody is bound to a fluorescent molecule. In indirect immunofluorescence, the primary antibody binds to the antigen but is not fluorescently labeled. A second antibody binds to the primary antibody and is fluorescently labeled.

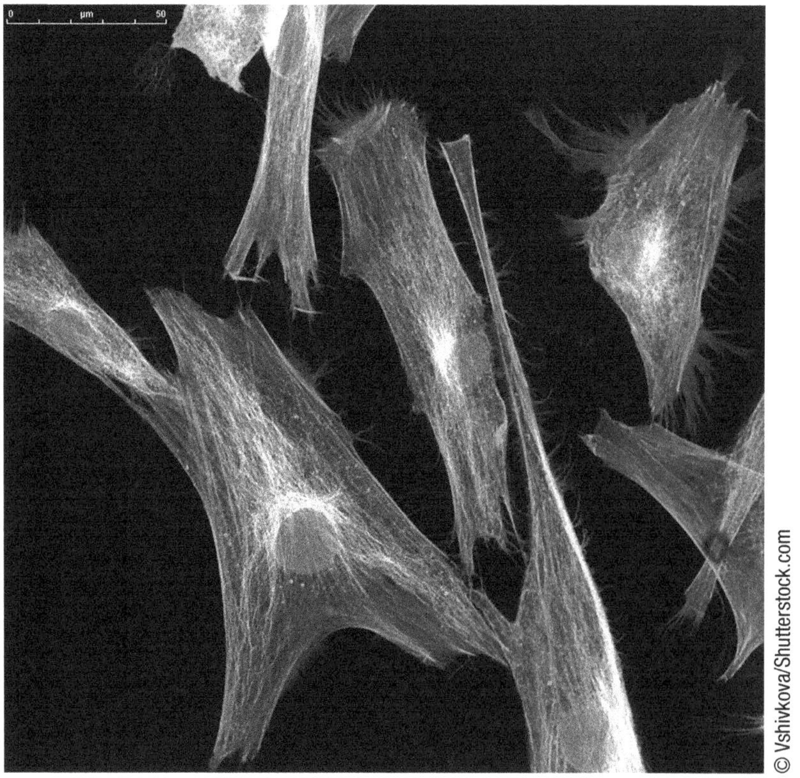

Figure 2.34 Immunofluorescence image. These cells were immunofluorescently stained for tubulin, and the image shows microtubules that are made of tubulin (green). The same cells were stained for actin (red) and for DNA (blue), using chemical dyes.

Binding and Bonding

3

1 Basic Chemistry Review: Binding and Bonding

This section is about how things happen chemically inside cells. It is about making and breaking chemical bonds, in particular those involved in binding reactions. For us the difference between binding and bonding is that binding involves formation of weak chemical bonds whereas bonding involves strong chemical bonds. It is also about energy and how the cells use the energy of ATP to make things go the way they want.

At its heart, this section is about binding of molecules. If you ask how do cells communicate, how do cells carry out their activities, and how do hormones work, the answer in general involves binding. Knowing the basics about binding, especially affinity, selectivity, and specificity of binding is important because it underlies just about everything cells do. It is also important from the standpoint of drugs and medicine because drugs typically work through binding. Our goal here is to start at a basic level and work our way to an understanding of binding and its relationship to how cells use energy.

A. Chemical Bonds

Chemical bonds are a stable attraction between atoms or molecules that depend on attraction of opposite charges or sharing of electrons. Atoms are the most stable when their outermost or valence shell of electrons is full (often this will be two or eight). They can achieve this state by gaining or losing electrons or by sharing electrons. When atoms gain or lose electrons they become ions, and the attraction between oppositely charged ions leads to the formation of ionic bonds. For example, in sodium metal, the atoms are uncharged. However, sodium has only one electron in its outer shell, and it readily loses this electron to become a sodium ion with a +1 charge (a cation). Chlorine is a gas where the chlorine atoms share electrons and are uncharged. Chlorine atoms in the uncharged state have seven electrons in the outer shell such that adding one more electron will fill the shell. When chlorine takes on an extra electron it becomes a chloride ion with a −1 charge (an anion). Sodium ions with their +1 charge are attracted to chloride ions with their −1 charge due their opposite charges giving rise to salt. There are other ways atoms can ionize. For example, a carboxylic acid can lose an H^+ to become a carboxylate anion, whereas an amine group can bind H^+ to become a positively charged ammonium ion.

At the other extreme there are atoms that usually bond by sharing electrons rather than losing or gaining them. These are known as covalent bonds. Sharing electrons accomplishes the same goal of obtaining a full outer shell but without becoming an ion. For example, the hydrogen atom is made up of a single proton and electron but its outermost shell needs two electrons to be full. Hydrogen exists as a gas where two hydrogens are bonded together by sharing their electrons.

The two atoms in a covalent bond can share electrons equally, or one of the atoms can tend to hog them. It depends how much attraction each atom in the bond has for the electrons, a property known as electronegativity. When the two atoms have similar electronegativities, the electrons are shared more or less equally, giving rise to a bond that is nonpolar. Examples of this are seen when two identical atoms such two hydrogens or two oxygens bond to each other to form H_2 or O_2. Since the two bonding atoms are identical, they share equally. The same can be said when hydrogen bonds to carbon. Although these atoms are not identical, they have similar electronegativities so the bonds are nonpolar. This is why hydrocarbons, those composed of carbon and hydrogen, are nonpolar and hydrophobic.

The situation is different when two atoms of greatly different electronegativities bond together. In this case, the bond is a polar covalent bond because the shared electrons spend more time around the more electronegative atom. This can be illustrated in the formation of water from oxygen and hydrogen gases. In the gas form, the H–H bonds and the O–O bonds are nonpolar. However, in water, where hydrogens are bonded to oxygen, oxygen is much more electronegative than hydrogen, so the electrons spend more time around the oxygen. The result is that the hydrogens are partially positive (δ+) whereas the oxygen is partially negative (δ–). Given that in water both hydrogens are positioned on the same side of the oxygen, the whole water molecule is a dipole, where one side is partially positive and the other partially negative. This feature gives water many of its remarkable properties.

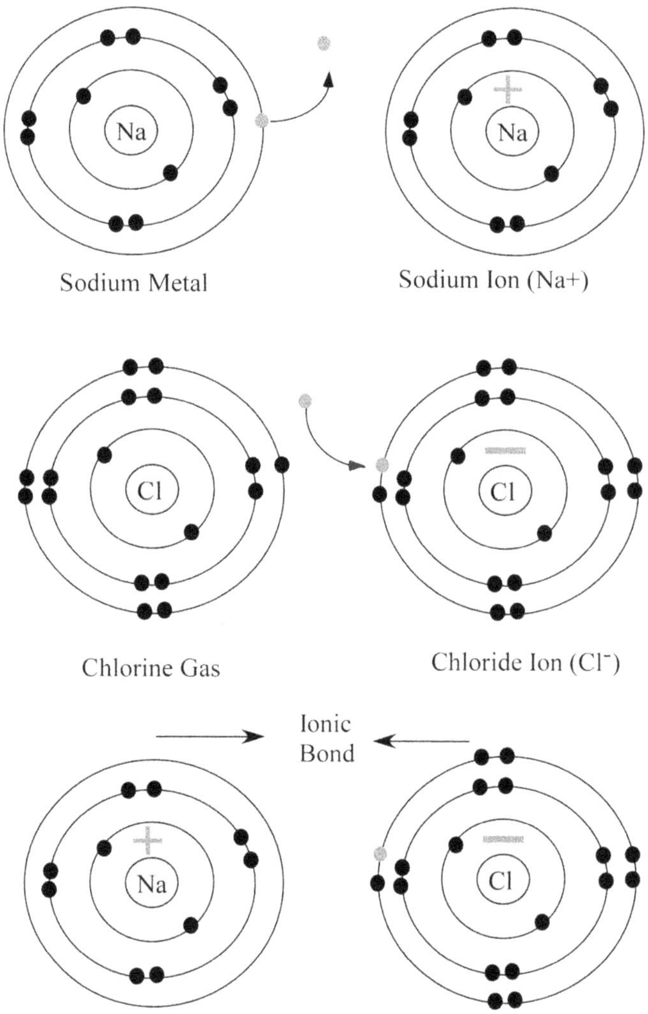

Figure 3.1. Ionic bonds' neutral atoms have the same number of protons and electrons, making them electrically neutral. But this is often not the most stable state of an atom. Atoms are most stable when they have a filled outer (valence) shell (typically two or eight electrons for most of the atoms we are studying). Atoms can obtain a filled outer shell by losing or gaining electrons. When this happens, a neutral atom will become charged, + or –, corresponding to whether the atom lost electrons (+) or gained electrons (–). Oppositely charged ions are attracted to each other, giving rise to ionic bonds.

Source: M. Poenie.

B. Strong and Weak Chemical Bonds

The strength of a chemical bond is defined by the energy needed to break it apart. Both covalent and ionic bonds can be strong. Covalent bonds are usually the strongest bonds holding cellular molecules together. Ionic bonds can be strong under certain circumstances, such as the calcium phosphate in bone. However, because water is a dipole, it can form weak bonds with ions surrounding them and keeping them apart. This effect of water greatly weakens ionic bonds. An example is seen in salt. Salt in crystal form is relatively strong, but when added to water the sodium and chloride ions come apart due to being surrounded by water molecules.

1) Weak Chemical Bonds

Weak chemical bonds, primarily hydrogen bonds and Van der Waals bonds, are exceptionally important in cell function. They are responsible for how proteins fold, how DNA strands hold together, how macromolecular cell structures are built, and how enzymes work. By "weak," we mean that individual weak bonds are easily broken. For example, individual hydrogen bonds or Van der Waals bonds are easily broken apart by the normal jostling about of water molecules.

a. Hydrogen Bonds

Hydrogen bonds form between a partially negative atom and a partially positive atom. One example is the hydrogen bonds that form between water molecules. In water, the more electronegative oxygen bears a partially negative charge whereas the hydrogens are partially positive. We should add that the hydrogens are not arranged symmetrically with respect to the oxygen but are shifted to the same side and are 107 degrees apart. The result is a molecule that is partially positive on one side and partially negative on the other. Water molecules form hydrogen bonds by the attraction of the partially negative oxygen to the partially positive hydrogens.

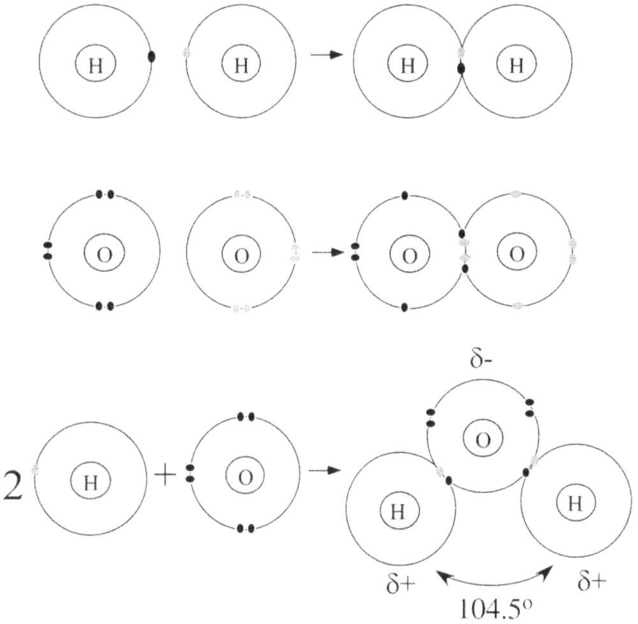

Figure 3.2. Covalent bonds: Another way that atoms can obtain a filled outer shell is by sharing electrons. A neutral hydrogen atom has only one electron but its outer shell takes two electrons to fill. Two hydrogen atoms can share their two electrons in a covalent bond. Oxygen has only six electrons but its outer shell holds eight. It achieves eight in the outer shell by having each oxygen share two electrons. Polar covalent bonds: Atoms have different strengths of pull on electrons (referred to as electronegativity). When oxygen shares with hydrogen, it hogs the electrons, so to speak. The result is that oxygen becomes partially negative, leaving hydrogen partially positive. This arrangement is referred to as a polar covalent bond.

Source: M. Poenie.

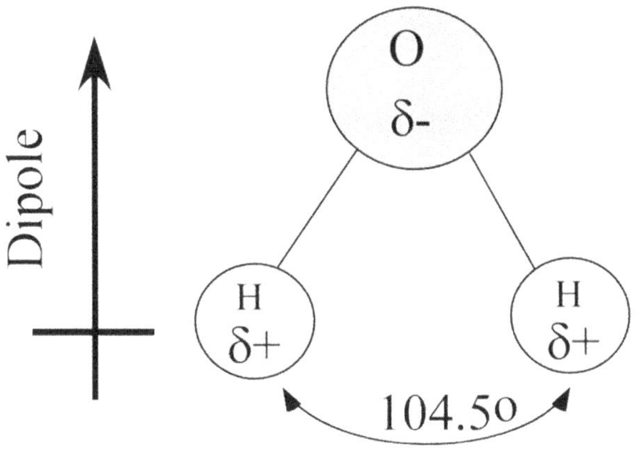

Figure 3.3. In water molecules two hydrogens are bound to an oxygen molecule. They hydrogens, being 104.5 degrees apart, are both on one side of the oxygen molecule. Since the hydrogens are partially positive and the oxygen is partially negative it is said that water molecules have a dipole moment and that water is a dipole. This just means that as one goes from the hydrogen side to the oxygen side the electric field goes from positive to negative.

Source: M. Poenie.

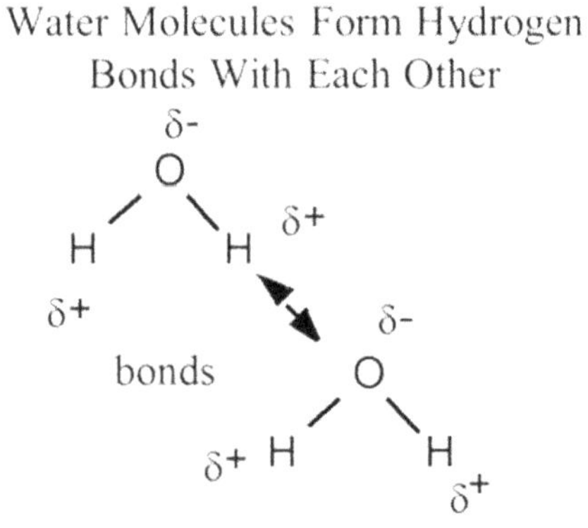

Figure 3.4. Hydrogen bonds Due to water being a dipole, water molecules want to orient so that the partially positive hydrogen atoms are close to a partially negative oxygen atom. This type of association between the hydrogen and oxygen is known as a hydrogen bond. Hydrogen bonding between water molecules account for a number of its unique properties.

Source: M. Poenie.

There are a variety of examples where a partially positive hydrogen (known as the donor) binds to a partially negative atom (the acceptor) and these are all called hydrogen bonds. Nitrogen is strongly electronegative and, like oxygen, it shares electrons unequally with hydrogen. In the case of an N–H bond, the nitrogen will be partially negative and the hydrogen will be partially positive. The partially positive hydrogen can then form a bond with a partially negative oxygen or a partially negative nitrogen.

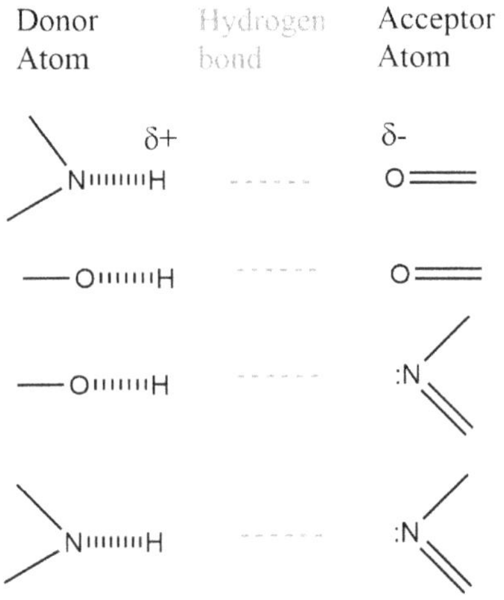

Figure 3.5. Various types of hydrogen bonds. In a hydrogen bond, the donor (hydrogen) is the partially positive member and the acceptor (oxygen or nitrogen) is the partially negative partner.

Source: M. Poenie.

b. Many Weak Hydrogen Bonds Combine to Make Strong Bonds Overall

Although an individual hydrogen bond is weak, many hydrogen bonds can work to bind molecules together strongly. An example is seen in an enzyme that binds to a trisaccharide molecule (red, below).

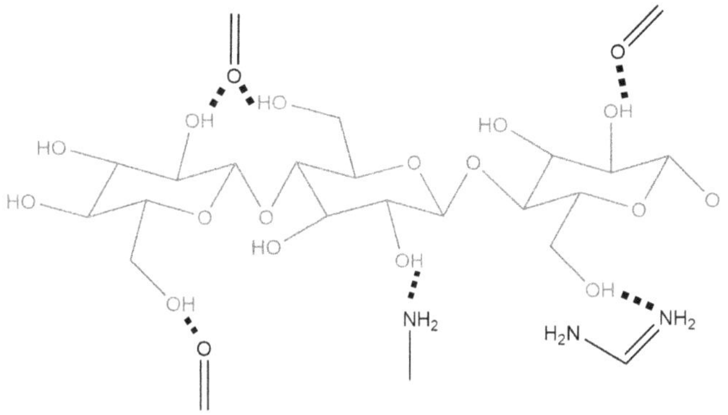

Figure 3.6. Individual hydrogen bonds are weak and break easily at room temperature. However binding involving many weak bonds can result in strong binding. In this figure, the binding of a trisaccharide (red) to an enzyme involves formation of many hydrogen bonds between the enzyme and the trisaccharide.

Source: M. Poenie.

c. Van der Waals Bonds: Another Type of Weak Bond

Van der Waals bonds are weaker than hydrogen bonds and arise due to electrical interactions that develop when two atoms come close together. As the two atoms come closer and closer, the electrons around the atoms can shift such that one side becomes more negative and the other side becomes more positive such that the electrically polarized atoms now form a bind. It should be noted that Van der Waals bonds can form between atoms that are not normally involved in polar bonds. As such, they can form between nonpolar molecules such as hydrocarbons.

In order to form Van der Waals bonds, the binding atoms must be just the right distance apart: too close and they repel; too far apart and no attraction is felt. This has an important implication when it comes to the binding of two molecules together, using numerous weak bonds.

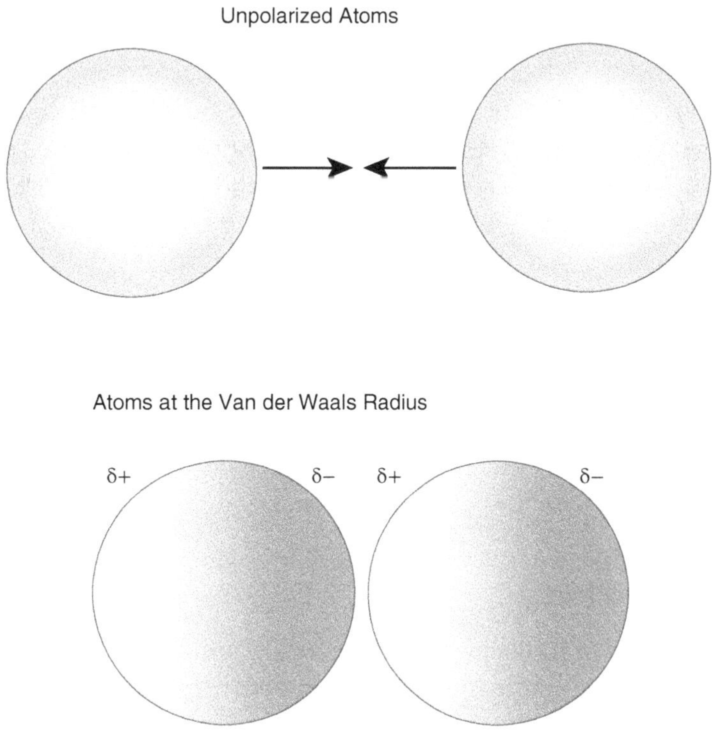

Figure 3.7.

Source: M. Poenie.

C. Binding

So far we have been concerned with chemical bonding. We now move to the idea of binding. "Binding," as we will use the term, concerns molecules that are stably associated due to the formation of weak chemical bonds. As mentioned earlier, one can obtain tight and stable associations between molecules when many weak bonds hold them together. These weak bonds can be hydrogen bonds, Van der Waals bonds, or ionic bonds. Since these weak bonds form only when the bonding atoms are close together and we are making many of them, it means that the bonding surfaces must be complementary. This introduces the idea of molecular fit. Like a hand and glove, the two molecules must fit together. However, fit alone is not enough. By fitting together, the two binding molecules much have bonding atoms placed in just the right places so that when the two binding partners come together, weak bonds will form.

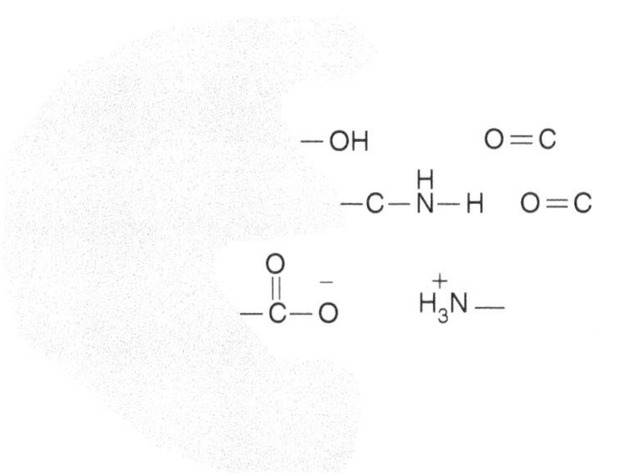

Figure 3.8. Binding between biological molecules often depends on the formation of weak chemical bonds. Here, two hydrogen bonds and an ionic bond are shown.

D. Hydrophobic Interactions

Hydrophobic molecules can bond together due to forming Van der Waals bonds. However, they can also associate due to the actions of other molecules, in particular, water. Water wants to bind to other water molecules. Adding drops of oil to water disrupts some of the bonds between water molecules and forces water to form an ordered cage around the droplet of oil. Both of these aspects are energetically unfavorable. Breaking bonds between water molecules requires energy, and formation of a cage around the oil drops involves decreasing entropy. Because these factors are energetically unfavorable, the oil droplets will be driven to coalesce so as to minimize the area of oil that is in contact with the water.

Hydrophobic interactions are important in cell biology. It is important in protein folding because hydrophobic regions of a protein will seek to coalesce in the interior of a protein to get away from water. Membrane lipids will also form structures that minimize the contact of water with the hydrophobic lipids.

E. Summary

Binding of molecules is due to weak bonds: hydrogen bonds, Van der Waals bonds, ionic bonds, and hydrophobic interactions. In order to bind together strongly, many weak bonds must be involved. Since bond distances are short, in order to make many weak bonds, the two binding surfaces must have complementary shapes.

2 Measurement of Binding

A binding reaction is a type of chemical reaction that arises from the association of two molecules due to the formation of weak chemical bonds. For our discussion we will use the terms "unoccupied" or "occupied" receptor [R] to stand for the free or bound states of one of the binding partners (e.g., a protein) and we will use the terms "free" or "bound" ligand [L] to refer to the other binding partner. The brackets [ ] refer to molar concentration.

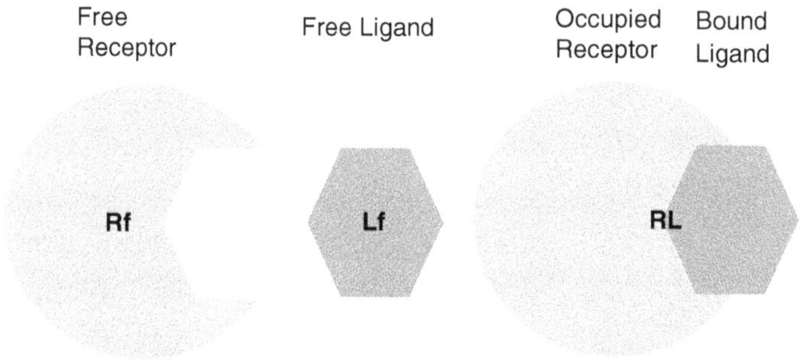

Figure 3.9. To illustrate a binding reaction we will use the term receptor, typically a large molecule like a protein, and the term ligand for a small molecule that binds to the protein. The receptor can be in two forms. It can either be free or unoccupied by ligand [Rf] or it can it can be occupied or bound to ligand in the ligand-receptor complex [RL]. Likewise the ligand can either be free [Lf] or bound to the receptor.

A. Ka (K_{eq}) and Kd Are Equilibrium Constants That Reflect the Strength of Binding

$$Ka = [RL]/[L_f][R_f]$$

$$Kd = [L_f][R_f]/[LR]$$

Where $[L_f]$ represents the molar concentration of free ligand and $[R_f]$ equals the molar concentration of free receptor, [RL] equals the molar concentration of the bound complex. For our discussion, we will call Ka* the association constant that relates the concentration of products/reactants when the reaction has come to equilibrium. You might say that Ka describes the tendency of free molecules to come together to form bound products. The strong tendency of molecules to come together and form products is described by a large Ka. The inverse of the Ka (= 1/Ka) is the Kd, or dissociation constant. Whereas Ka describes the tendency of reactions to come together to form products, Kd describes the tendency for bound molecules (products) to come apart and give free molecules. Ka and Kd thus look at reaction from opposite directions. Strong or high-affinity binding would equally entail a large Ka and a correspondingly small number for Kd.

1) Understanding Equilibrium Constants

In order to get a better idea of what Ka or Kd means, we need to think of what happens during a chemical reaction, and in particular, during a binding reaction. Let's start out with a lot of free ligand and unoccupied receptor and virtually no product (bound receptor ligand complex) Figure 3.9. All chemical reactions are in principle reversible, but when there is no product the reaction starts out going essentially one way. As the reaction continues, the concentration of reactants decreases and the speed of the forward reaction gets slower. Meanwhile we start building up products, and the rate of the reverse reaction increases. This process continues until the rate of the forward reaction equals the rate of the reverse reaction. At that point, we are at equilibrium and the concentration of the reactants (free ligand, unoccupied receptor) and products (receptor–ligand complex) are constant.

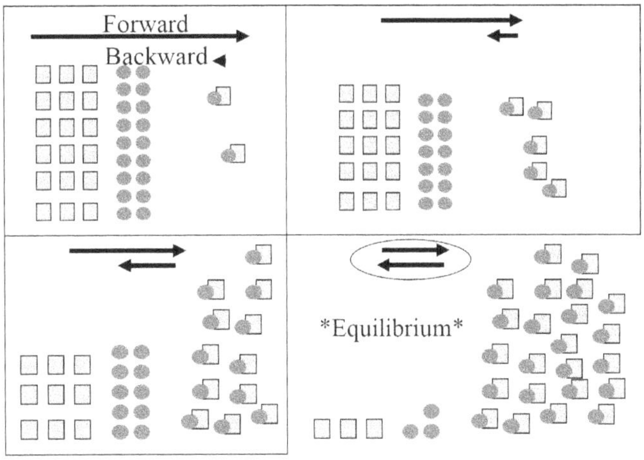

Figure 3.10. Progress of a reaction to equilibrium. The binding reaction begins with free receptor (boxes) and free ligand (circles). As the reaction proceeds, ligand and receptor bind to each other and the concentration of free ligand and receptor is reduced. Consequently, the rate of the forward reaction slows. Meanwhile, as the receptor and ligand bind, the concentration of the bound receptor-ligand complex increases and with it the rate of the backward reaction. These trends continue until the rate of the forward reaction equals the rate of the backward reaction, at which point, the system is at equilibrium.

Source: M. Poenie.

---

* Terminology for these constants can be confusing. Both Ka and Kd are equilibrium constants where Ka = 1/Kd). However Ka is normally called K_{eq} (The equilibrium constant) because Ka in other contexts is often used specifically with acids.

2) On and Off Rate Constants

We can describe the reaction in Figure 3.9 using on and off rate constants. The on rate constant (K^{+1}) is a number that when multiplied by the concentration of reactants gives you the rate going forward. The reverse or backward rate constant (K^{-1}) is a number that when multiplied by the concentration of products gives you the reverse rate. By definition, at equilibrium, the rate forward equals the rate backward.

Therefore, *at equilibrium*,

$$K^{+1} \times [\text{concentration of reactants}] = K^{-1} \times [\text{concentration of products}].$$

$$\text{Remember for a simple binding reaction Kd} = [\text{the concentration of reactants}] / [\text{concentration of products}]$$

Doing some simple math, we can see that at equilibrium

$$K^{+1} \times [\text{concentration of reactants}] = K^{-1} \times [\text{concentration of products}];$$

$$\frac{[\text{concentration of reactants}]}{[\text{concentration of products}]} = \frac{K^{-1}}{K^{+1}} = \text{Kd}.$$

Rate constants let us see the reaction from a little different perspective. For a molecule that has a low Kd (strong binding), the binding partners will come together quickly, such that K^{+1} is a large number. By the same token, once bound, the complex will come apart slowly, such that K^{-1} is a small number.

B. Binding Curves

We want to now show how binding varies as a function of concentration. First we need to say that at a high enough concentration of free ligand and receptors, everything will bind. Needless to say, at zero concentration nothing binds because there is nothing to bind. If we vary concentration from zero to infinity, somewhere between those extremes we will see things begin to bind (see Figure 3.11). For simplicity, we will assume that the receptor is a protein and its concentration is fixed. The ligand is a small molecule that binds to the receptor and whose concentration we can easily vary. As can be seen in Figure 3.1, Kd is also the free ligand concentration where 50% of the receptors are bound to ligand. The strength of binding between receptor and ligand is reflected in the concentration of ligand needed for binding to occur. Stronger binding between receptor and ligand means that the complex will form at low concentrations of ligand. Were the binding weaker, it would take higher concentrations of ligand to get them to bind. We can represent the strength of binding in terms of Kd. The *lower* the Kd, the lower the concentration of ligand needed to get binding and thus, the stronger the binding.

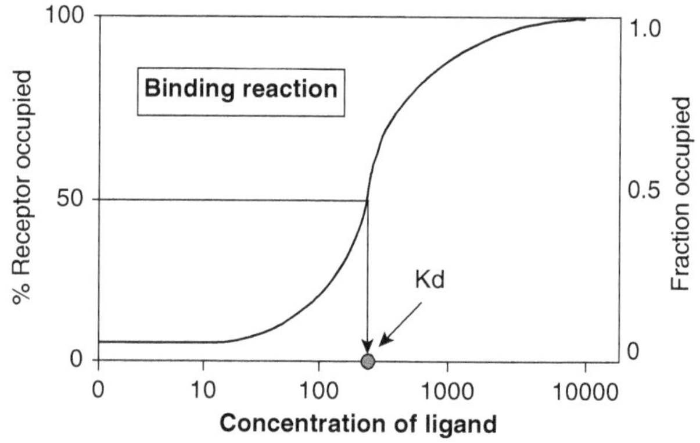

Figure 3.11.

C. Describing Binding in Terms of Free Energy

Binding reactions are chemical reactions, and chemical reactions are driven by energy. Chemical reactions occur because the free energy content of the reactants ($G_{reactants}$) is higher than that of the $G_{products}$. Delta G (ΔG) is the difference between the two ($\Delta G = G_{products} - G_{reactants}$). We said we measure the strength of binding in terms of concentration. Kd is an *equilibrium concentration.* Its units are in terms of molarity. However, we can easily convert between Ka or Kd and the standard free energy ($\Delta G°'$) of a reaction. We need to note that *standard free energy is the free energy of a reaction measured under standard conditions* where all the reactants and products start out at 1 molar (M) concentration and where T (the absolute temperature in degrees Kelvin) is 298K (25°C) and R is the gas constant. Chemists like to run their reactions at 1M acid, which is hard on biological molecules, so for biology we use $\Delta G°'$ (delta G naught prime) to designate that the reactions are run at a pH of 7 instead of 1M acid: $\Delta G°'$ is then a measure of how far equilibrium is away from the starting (standard) conditions. Standard free energy ($\Delta G°'$) is different from ΔG in that ΔG takes into account the actual starting conditions of the reaction – what the real concentration of reactants and products are.

$$\text{Standard free energy } \Delta G°' = -RT \ln K_{eq}$$

The important conclusion here is that <u>$\Delta G°'$, Kd, Ka, and rate constants</u> <u>are all connected</u>. Strength of binding can be measured in terms of Kd but this is also related to $\Delta G°'$ (or ΔG). As we will see, these ideas are essential for understanding how cells get things done. Put simply, cells use energy to change affinity and binding relationships.

Table 3.1. Values of $\Delta G°'$ for Some Values of K_{eq}

K_{eq}	$\Delta G°'$ (cal/mol)*
0.001	+4086
0.01	+2724
0.1	+1362
1.0	0
10	−1362
100	−2724
1000	−4086

Note. **Calculated from the formula* $\Delta G°' = -2.3\ RT \log K_{eq}$

What this chart shows us is that as Ka ($= K_{eq}$) gets larger – meaning that we end up with a larger and larger ratio of bound product to free ligands – there is a greater negative free energy associated with the binding reaction. This is a key concept and understanding it is absolutely necessary to get a good grip on this for the next section.

1) A Sample Calculation

Here is an example of a calculation where the K_{eq} is 4.692×10^3:

$\Delta G^{o\prime} = -2.3RT \log_{10} K_{eq}$ (note here log n is converted to log 10 by multiplying by 2.3)

$\Delta G^{o\prime} = -2.3(8.3\ J/mol/K)(298\ K)\log_{10}(4.692 \times 10^3)$

$\Delta G^{o\prime} = -20{,}900\ J/mol = -20.9\ kJ/mol$

2) Summary

$\Delta G = G_{products} - G_{reactants}$. In order for a reaction to go, $G_{reactants}$ must be greater than $G_{products}$. Therefore, in order for a reaction to go forward as written, ΔG must be negative.

ΔG^o is the free energy of a reaction run under standard conditions. $\Delta G^{o\prime}$ is the standard free energy with the added constraint that the reaction is run at pH7 rather than 1M acid.

$\Delta G^{o\prime}$ can be converted to Ka or Kd. Things bind together because the $\Delta G^{o\prime}$ for the binding reaction is negative. The more tightly they bind, the lower the Kd and the more negative the free energy of binding.

D. Binding and Unbinding

Remember, a reaction will take place (go forward as written) only if the free energy (ΔG) is negative. If the free energy for a reaction is positive as written, then the reaction simply goes backward from the way it is written. If there is a large negative free energy associated with free molecules binding to each other to form a complex, what should be true about the energy it takes to break apart the complex back into free molecules? If you answered that it will take are large input of free energy to break them apart you would be correct.

In a chemical reaction such as a binding reaction, the reaction will always proceed to equilibrium, at which point the free energy of the reaction is 0. In fact, the free energy for a chemical reaction describes how far away we are at the beginning from that equilibrium point where the free energy is 0. If the free energy for a reaction is negative, then we proceed downhill until we reach the bottom (equilibrium). If the free energy for the reaction is the same amount but positive, then we are just as far away from equilibrium, but we go in the opposite direction to get to equilibrium Figure 3.12.

All this is background to get us to the real cell biological point we want to discuss next.

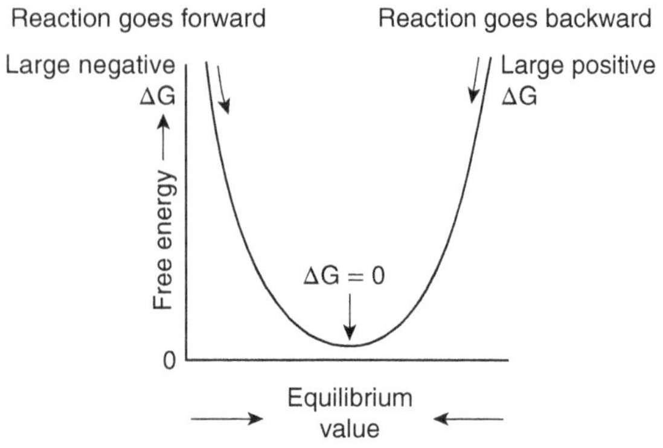

Figure 3.12. The ΔG always proceeds to 0 In a reaction such as a binding reaction, the reaction will go forward if there is a negative ΔG for binding. At the reaction proceeds, the ΔG decreases until it reaches zero at which point no further reaction can take place. If the ΔG were positive, the reaction would go in the opposite direction until the ΔG was 0.

3 ATP as the Energy Currency of the Cell

Perhaps you are wondering what all this has to do with cell biology. What we have covered so far are principles that are basically true of all chemical reactions. Of particular focus here are binding reactions where one molecule (let's say a protein) binds to another molecule (which could be another protein or some small molecule).

A. ATP and the Concept of Coupling

When a reaction as written has a positive free energy, it will not go forward unless there is an added source of energy to drive it forward. What is needed to make the reaction go forward is to couple an energetic reaction with a large negative ΔG energy to the reaction that we want to go forward. The energy released by one reaction has to have a sufficiently negative ΔG so that the overall ΔG for both reactions is negative. Often, the molecule that the cell uses to drive other reactions forward is adenosine triphosphate, or ATP. ATP is a "high energy" molecule that is often called the energy currency of the cell. Our goal here is to gain an idea of how the cell uses the energy of ATP to get things done.

ATP is a high-energy molecule because it has three phosphates linked together in what are called phosphoanhydride linkages. Putting three phosphates together like this is chemically unstable. We say that ATP has a large negative standard free energy of hydrolysis ($\Delta G = -30.5$ kJ mol^{-1}). When ATP is hydrolyzed in a reaction with water, we end up with ADP and free phosphate (inorganic phosphate, or Pi). But that alone does not get us anything. We need some reaction of ATP to participate in or be coupled to the reaction.

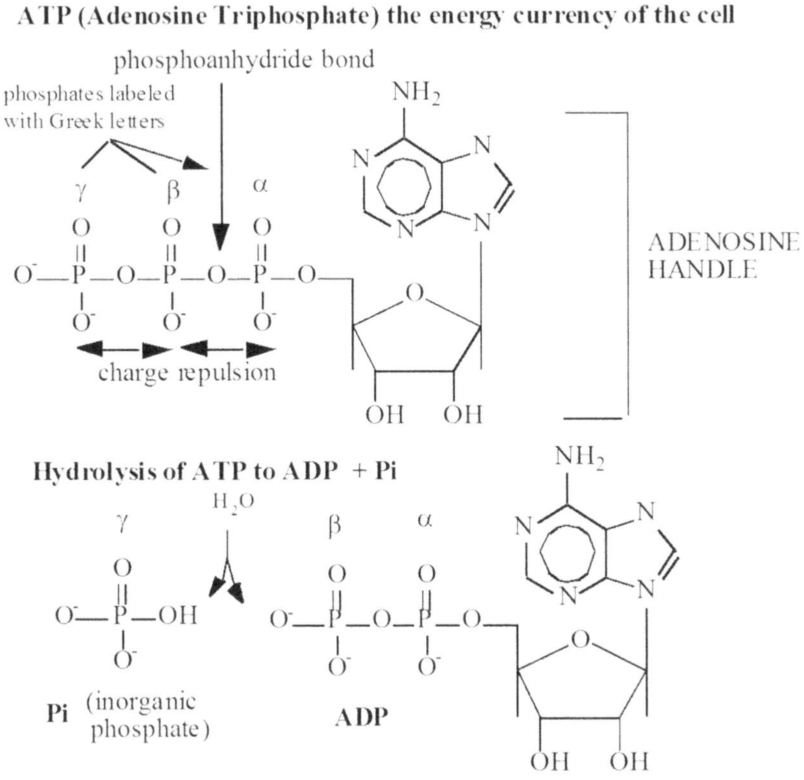

Figure 3.13. ATP the energy currency of the cell. ATP is a high energy molecule that is use to drive reactions in the cell. The reason it is a high energy molecule is due to the three phosphates linked together in phosphoanhydride linkages. The phosphates want to come apart due to both the charge repulsion between the negative charges on the phosphate groups and the increase in entropy that occurs when the phosphate groups come apart.

Source: M. Poenie.

In many cases, ATP gets involved in a reaction either through the binding of ATP or by transferring one of the phosphates from ATP to another molecule. This phosphate may ultimately be removed so that overall, we end up with ADP and P, but in the interim, adding a phosphate to another molecule like a protein did some work. It changed the protein to a different state, or conformation.

Now, at last we can get to the key point of all this. If we ask how cells use energy to get things done, we can say, for many cases, *cells use energy, especially the energy of ATP, to change the binding affinity* of components involved in a binding reaction.

B. Cells Use Energy to Change Affinity

A huge part of cell biology concerns the binding and unbinding of molecules. This is how a cell directs many of its activities. Consider the transport of molecules across the membrane against a concentration gradient. As written, this process would have a positive ΔG. This process, known as active transport, involves a transporter (a receptor) binding to a substance (ligand) at low concentration on one side of the membrane and releasing it on the other side of the membrane, where the concentration is higher. So there are two binding reactions, one which has a lower Kd (high-affinity binding, because the concentration of ligand is low) and the other which must have a higher Kd (low affinity) so that it will release the ligand on the other side of the membrane, where the ligand concentration is already high.

1) Key Concept

If a binding reaction has a strong tendency to go toward binding, then it will be hard to get the ligand off the receptor. In fact, since in high affinity binding energy is released when a ligand and receptor bind, it will take an input of energy to get the ligand and receptor apart. The cell has many ingenious ways to implement this principle. For example, suppose ligand (L) and receptor (R) come together in a strong binding reaction. You could use the energy of ATP to get them apart. One way is to phosphorylate the ligand or receptor so that L and R do not bind anymore.

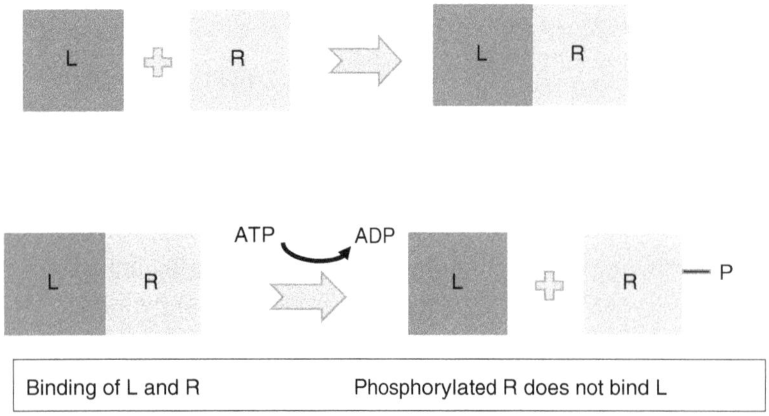

Figure 3.14. Here the energy of ATP is used to cause L and R to unbind because the phosphorylation of R decreases its affinity for the ligand

From our previous discussion, it should be clear that Kd is directly related to free energy. An essential idea here is that *we cannot change the Kd of a binding reaction without changing the free energy of the binding reaction.* Therefore, if we want to change Kd to go from strong binding to weak binding, we will have to put in the corresponding amount of energy to change the free energy of the binding reaction, so that overall the reaction still has a negative free energy.

We can illustrate this with an important cellular process called active transport, where a solute (red dots) is transported from low concentrations on one side of the membrane to higher concentrations on the other side of the membrane (Figure 3.15). As described, this process would

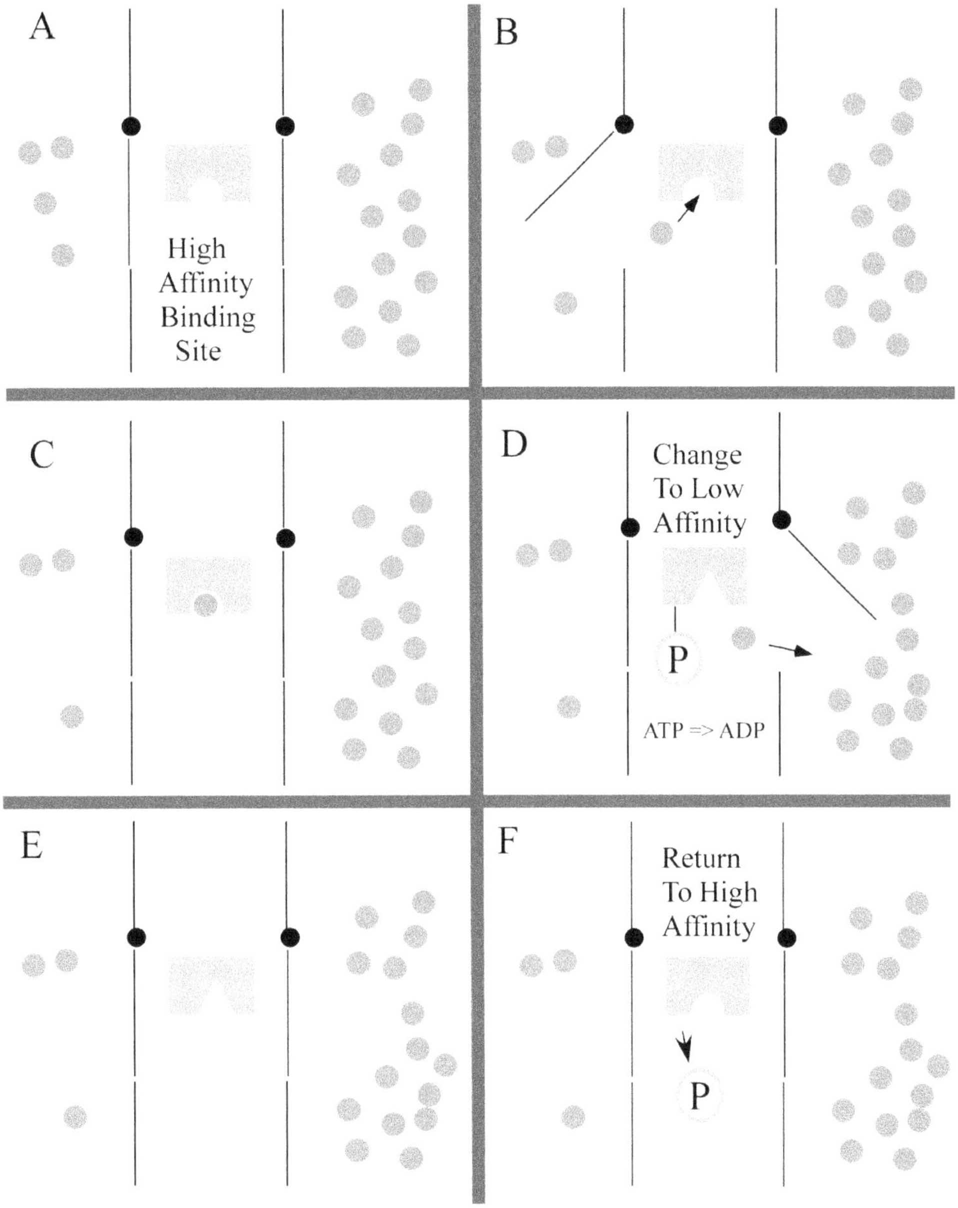

Figure 3.15. Active transporters often use double gating combined with a change in affinity. In this model we start with a high-affinity binding site (A). B. When the first or left gate (or door) opens to the low concentration side of the membrane, the high-affinity binding site will bind the substrate. C. Then the gate closes and D the door to the high concentration site opens while the binding site itself converts to low-affinity binding using the energy provided from ATP. For this example we assume that ATP is used to phosphorylate the transporter. Under these conditions the substrate will diffuse to the high-concentration side. E Then the door closes and the binding site reverts to high affinity by removing the phosphate. The change in affinity requires an input of energy, which accounts for the energy required for active transport.

Source: M. Poenie.

have a positive ΔG because the solute really wants to go in the opposite direction (from higher concentration to lower concentration). In order to move things from lower concentration to higher concentration, the cell will need to put in some energy so that the overall ΔG for the transport reaction is negative. So, what we want to focus in on here is how the cell accomplishes this. How does it put in energy to drive the transport energetically uphill? The answer is through changing affinity of the binding reactions.

In active transport, the transport system can be envisioned as a passage way with two gates or doors, one on the low-concentration side and one on the high-concentration side. During transport, the door on the low concentration side (door 1) opens, and the solute binds with high affinity to the transporter (inside the passageway). Door 1 then closes, and the door on the high-concentration side (door 2) opens. Now the problem here is that the transporter is binding to the solute too tightly to let go, especially in the presence of high concentrations of solute. So, what needs to happen is that the binding affinity needs to be reduced (i.e. Kd for binding needs to increase) so that the transporter will let go of the solute. This requires an input of energy. This energy could be in the form of ATP that phosphorylates the transporter and reduces its affinity. The transporter then lets go of the solute which diffuses away and door 2 closes. The transporter now has to get rid of the phosphate that was attached to it so that it can revert to the high binding affinity state. Since both the transfer of phosphate from ATP to the transporter and the removal of the phosphate are energetically downhill, these steps both help drive the transport reaction.

Overall we see that for active transport there are two Kds for the transport reaction, Kd1 and Kd2. The energy for transport is related to the difference in free energy for the two Kds. The cell transporter uses the energy of ATP to change affinity, and that is what drove the transport.

2) Conclusion

ATP is the energy currency of the cell. Other ribonucleotides (GTP, CTP, UTP) could in principle work just as well, but ATP is most commonly used to drive reactions in the cell. The question we tried to answer here is how cells use ATP to get things done. The answer in many cases is that *the energy of ATP is used to change affinity*. We will see many examples of this throughout the course.

Proteins

4

1 Amino Acids, the Building Blocks of Proteins

Proteins are molecules made from twenty different L-amino acids linked together by peptide bonds. Each amino acid differs from the others by its R group. The amino acids used in proteins are referred to as alpha amino acids (Figure 4.1). From an old convention, the first carbon next to the carboxyl carbon is called the alpha carbon. An alpha amino acid has the amine (NH_2) connected to the alpha carbon. An R group is also attached to the alpha carbon. The R group is what distinguishes the different amino acids. Amino acids are joined together by a reaction between the carboxyl of one amino acid and the amine group of another. The overall reaction can be considered a dehydration reaction, since the bond is formed with the removal of water. The resulting bond is called a peptide bond.

Amino acids can exist in mirror image forms referred to as L and D. During protein synthesis, only L amino acids are used to make the protein. D amino acids can be found in proteins, but this is due to the conversion of an L amino acid to the D form after the protein is made.

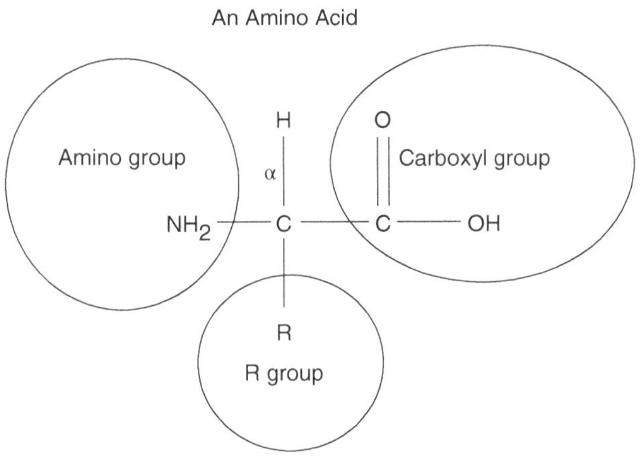

Figure 4.1. Amino acids used in proteins all have the same basic structure, except for the R group. There is an amino group attached to the second carbon from the carboxyl group. This is known as the alpha carbon, and amino acids are often called alpha amino acids. Amino groups are bases when free. Also attached to the alpha carbon is a carboxyl group. Carboxyl groups are acidic, so they are often called carboxylic acids. There are twenty different amino acids used to make proteins, and what distinguishes them is that each has a different R group. The various R groups give the twenty different amino acids their different chemical properties.

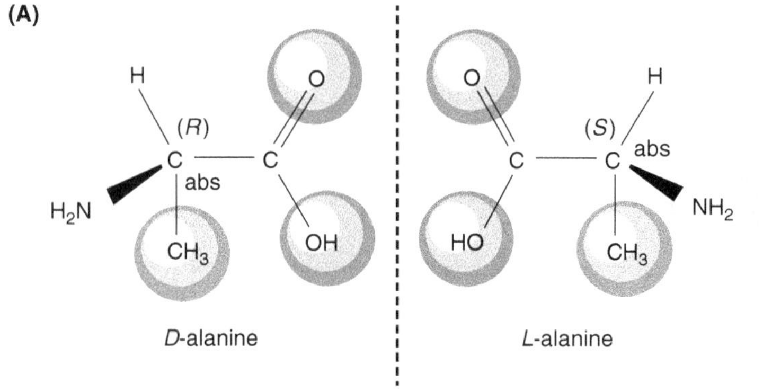

Figure 4.2. All the amino acids used in the proteins we make come in two forms (D and L) that are mirror images of each other. Proteins are synthesized using only L amino acids, although, after the protein is made, there are some cases where an L amino acid is converted to its D counterpart.

A: © Kendall Hunt Publishing Company; B: XiXinXing/Shutterstock.com.

A. R Groups

The twenty different amino acids used to be described by a three-letter designation (e.g., Ser, Ala, Thr) but nowadays they are designated by a single letter. If you want to read the literature, it will be necessary to know the single-letter designation. The twenty different amino acids have different properties due the properties of the R group and can be broadly grouped into four categories: hydrophobic, acidic, basic, and polar. Hydrophobic amino acids have a hydrocarbon chain as the R group. Hydrophobic groups avoid being in water if they can.

Polar amino acids (e.g., S, T, Y) have R groups that make them more polar. This could include the acidic and basic amino acids, but we will separate those out into individual categories. For example, serine (S), threonine (T), and tyrosine (Y) each have an OH group that makes them polar.

Acidic amino acids (D and E) have a carboxyl group as part of their R group. This carboxyl group ionizes (loses an H^+) such that it becomes negatively charged. The three basic amino acids (K, R, H) have R groups that ionize by binding to an H^+ ion at neutral pH and becoming positively charged.

Acidic and basic amino acids are known as weak acids and bases, because they do not ionize completely in water. By contrast, strong acids and bases ionize completely in water. Weak acids and bases vary in their degree of ionization as a function of pH. We characterize weak acids and bases by their pKa, which is the pH where 50% of the compound is ionized and 50% is not ionized. For carboxyl groups, the pKa is roughly around pH 5. For amine groups, the pKa is roughly around a pH of 9. The relative amount of the ionized and nonionized forms changes as the pH goes above or below the pKa. For a particular carboxyl group on a population of proteins, as the pH becomes more acidic (i.e., the pH goes lower) more and more of the carboxyl groups will become protonated. The reverse happens as you get more basic. For amine groups it is the opposite. As you get more acidic, a particular amine group on a population of proteins will more often become protonated (positively charged).

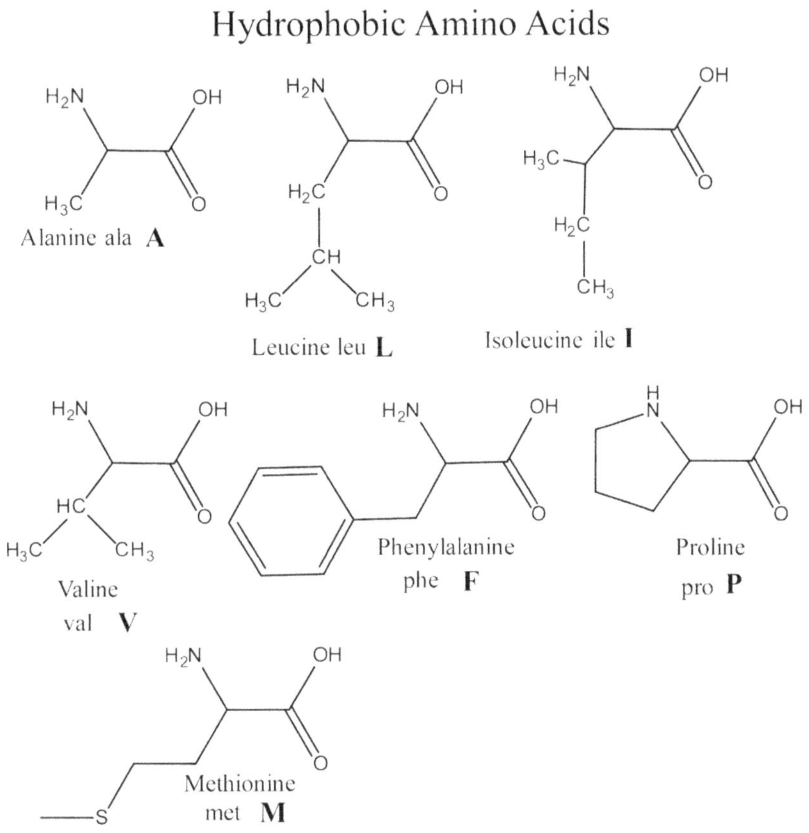

Figure 4.3.

Source: M. Poenie

B. Charges on Amino Acids as a Function of pH

As mentioned above, amino acids have different charges, depending on the pH. In strong acid the carboxyl group is protonated to give a 0 charge and the amino group is protonated to give a positive charge. At this point the amino acid is net positive. As the pH is raised toward neutral (pH 7), the carboxyl group will ionize (give off its hydrogen ion) to become negative. At this point, the amino acid carries both a positive and a negative charge and has no net charge. This is called a zwitterion. When the pH is raised further (above 9), the amino group loses its hydrogen ion to become neutral, so that the amino acid has a net negative charge.

$COOH^0$	$COO^-/COOH^0$	COO^-
pH 0	pH 5	pH 14
Nonionized	← Carboxyl pKa →	Completely ionized
	50% ionized	
$R\text{-}NH_3^+$	$R\text{-}NH_3^+/R\text{-}NH_2^0$	$R\text{-}NH_2^0$
pH 0	pH 9	pH 14
Completely ionized	← Amine pKa →	Completely nonionized
	50% ionized	

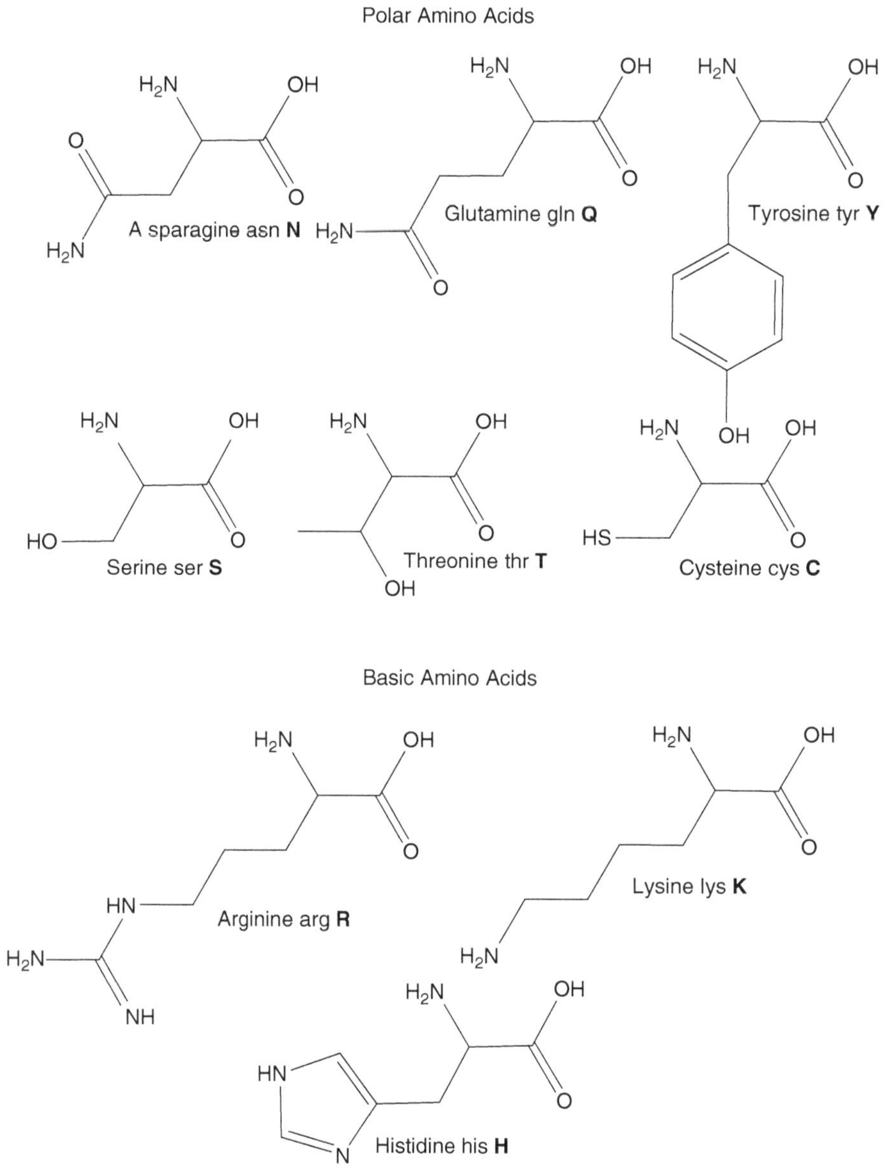

Figure 4.4.

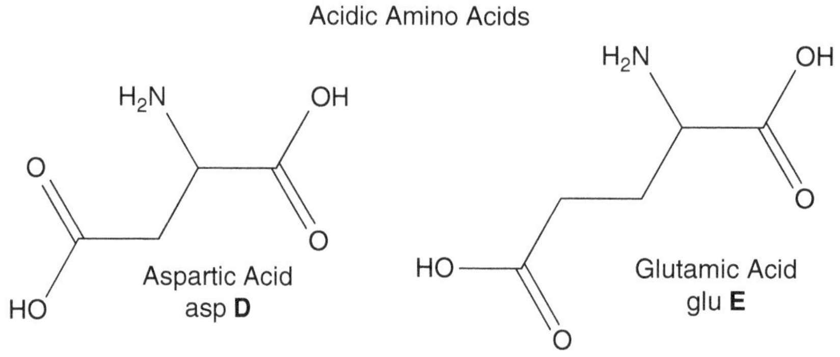

Figure 4.5.

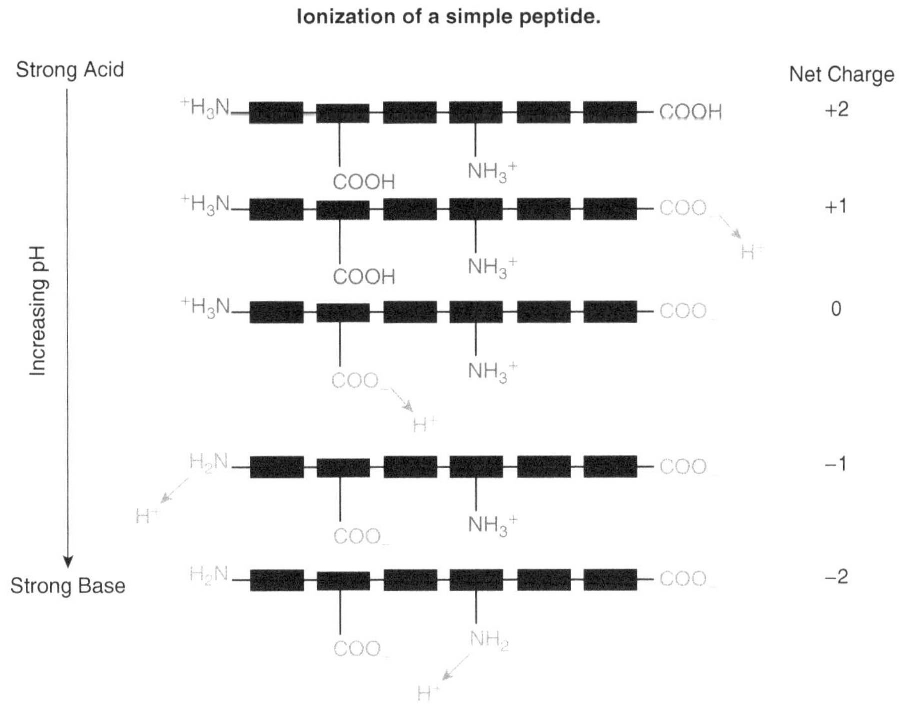

Figure 4.6. The net charge on peptides and proteins.

Peptides behave similarly to amino acid, as the pH moves from acidic to basic. However, peptides might include acidic and basic amino acids. In Figure 4.6 we show a peptide with one basic and one acidic amino acid in the chain and how the net charge changes when that peptide moves from an acidic to a basic environment.

1) Changing the pH Environment of a Peptide

Here we start at the top of Figure 4.6, with a peptide containing one acidic and one basic amino acid. In addition, the peptide has a free amino group coming off the first amino acid (N terminus) and a carboxyl coming off the last amino acid (C terminus). We start out with the peptide in a strongly acid solution. In acid, all the amino groups are protonated, so they have a positive charge and all the carboxyls are also protonated, so they have 0 charge. In this state, the charge on the peptide is +2. As we now increase the pH, we reach a point where the first and then second carboxyl groups lose an H^+. With loss of H^+ the net charge on the peptide goes first to +1 and then with the loss of the second H^+ the net charge on the peptide goes to 0. With increasing pH, eventually the first amino group will also lose an H^+, making the net charge −1, and with the loss of the last H^+ the net charge goes to −2. Note that the pH where the protein has no net charge is known as the *isoelectric point*. The isoelectric points of proteins vary due to the number of basic and acidic amino acids that are present in the protein.

C. Isoelectric Focusing: An Application of the Understanding of pH and Protein Net Charge

The relationship between pH and the net charge on a protein is the foundation for a protein separation technique called isoelectric focusing. In isoelectric focusing, one has a tube of acrylamide gel containing ampholytes. The ampholytes set up a stable pH gradient when one end of the tube is placed in acid and the other end is in base and electricity is applied. A solution of protein is added to one end of the tube and a voltage is applied so that the positive electrode is at the acid end and the negative electrode is at the minus end. Proteins at the acidic end are positively charged and move toward the opposite end. As they move, the pH becomes gradually more basic. As the environment becomes more basic, proteins will lose hydrogen ions (positive charges). Eventually the protein reaches a pH where its net charge is zero. This is known as the isoelectric

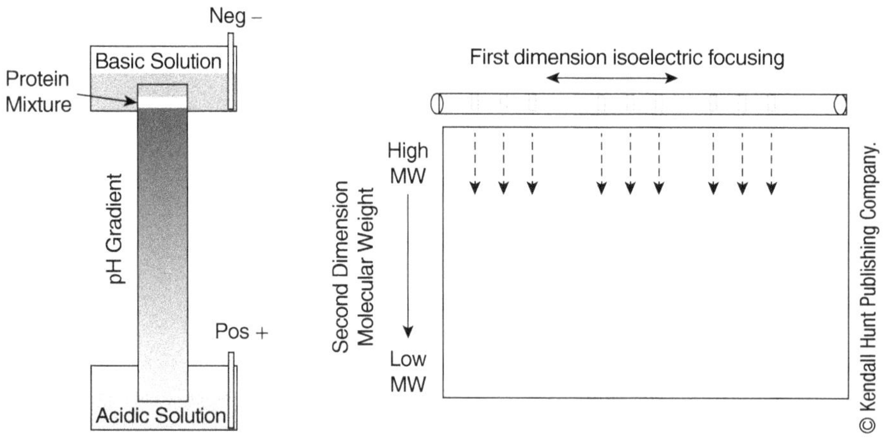

Figure 4.7. (**A**) In isoelectric focusing, a tube is filled with an acrylamide gel containing ampholytes. When one end of the tube is placed in acid and the other in base and a voltage is applied, the ampholytes move to different parts of the gel and establish a stable pH gradient from one end to the other. When a mixture of proteins is then added to one end and the voltage reapplied, proteins will move toward the opposite end, but they will stop moving when they reach the pH in the gel where they have no net charge. (**B**) In the second dimension, the tube containing the proteins separated by isolelectric focusing is placed on an SDS gel slab (see Section 5.A), which separates proteins by their molecular weights.

point, and it is at that pH where the protein stops moving. Different proteins have different isoelectric points and will stop at different places in the tube. Isoelectric focusing is the first dimension in 2-dimensional gel electrophoresis and the second dimension, SDS polyacrylamide gel electrophoresis, separates proteins by molecular weight.

D. Amino Acids of Special Note

1) Serine, Threonine, and Tyrosine

Serine threonine and tyrosine are significant because they are often implicated in catalytic functions or are often phosphorylated by *kinases*. Serine, threonine, and tyrosine each have an OH as part of their R group that can be phosphorylated.

2) Cysteine

Cysteine is special because it has an SH (sulfhydryl) group. The sulfhydryl can be oxidized to a disulfide bond (a covalent bond). Formation of disulfide bonds is reversible and involves reduction, the reverse of oxidation. Inside the cell,the cytoplasm is a reducing environment whereas the environment outside the cell or inside the endoplasmic reticulum is oxidizing. Formation of disulfide bonds is an important means for stabilizing the structure of extracellular proteins. On the other hand, for proteins that have been isolated from the cytoplasm, it is often important to prevent their cysteines from being oxidized. To prevent oxidation, biochemists use reducing agents to keep cysteines in the reduced form. Examples of these reducing agents are dithiothreitol and beta-mercaptoethanol.

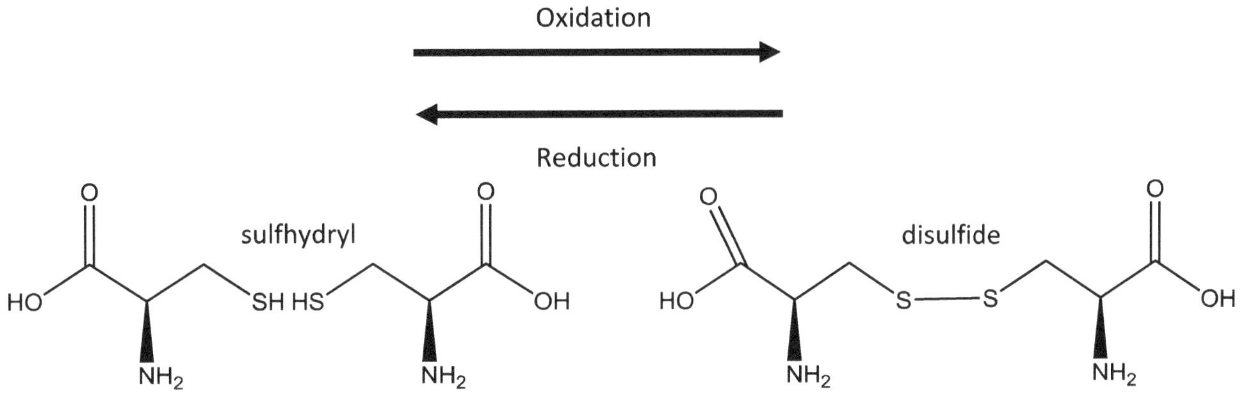

Figure 4.8. The sulfydryls (SHs) of two cysteine amino acids can be joined together to form a disulfide bond (S—S) by a process known as oxidation. The disulfide bond is a strong covalent bond that stabilizes the structure of extracellular proteins. Inside the cell the environment is reducing, such that most of the time, cysteines are reduced (in the SH form). Examples of extracellular proteins that are stabilized by disulfide bonds are antibodies, insulin, and keratin molecules in hair.

Source: M. Poenie.

You might be familiar with reducing and oxidizing disulfide bonds in a hair treatment known as a permanent. Hair is made up of many chains of keratin proteins that are bonded together by disulfide bonds. These disulfide bonds glue the keratin molecules together in particular forms, depending on the type of hair (straight, curly, etc.). In a permanent, the hair is first treated with a reducing agent that reduces the disulfide bonds to individual sulfydryl groups. This allows the keratin molecules to be rearranged to a different pattern. Once the desired pattern has been achieved, the hair is treated with an oxidizing agent that reforms the disulfide bonds and locks the keratin molecules into the new arrangement.

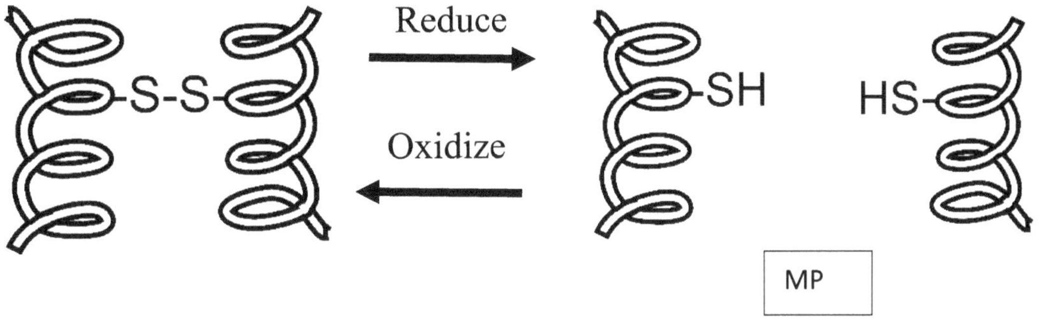

Figure 4.9. The disulfide bonds of keratin molecules in hair can be reversibly reduced or oxidized.

Source: M. Poenie.

2 Protein Structure

Protein structure is described in terms of different levels; primary, secondary, tertiary, and quaternary.

A. Primary Structure

Primary structure is the covalently bonded structure of a protein. It is the sequence of amino acids starting from the N-terminus (amino acid with free amino group) and reading to the C terminal amino acid (the amino acid with a free carboxyl group). It also includes the positions of any disulfide bonds.

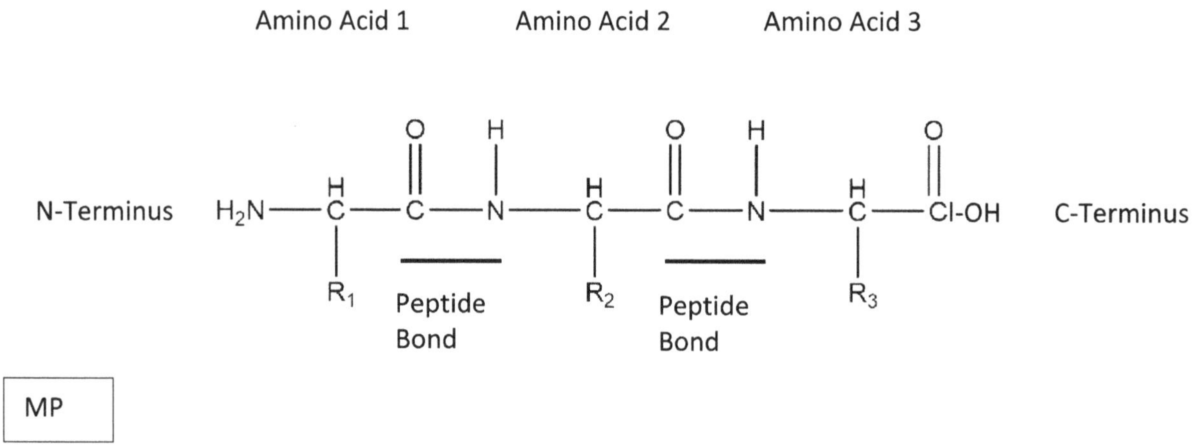

Figure 4.10. Primary structure is the sequence of amino acids starting from the N-terminus and reading to the C terminus.

Source: M. Poenie.

B. Secondary Structure

Secondary structure is the hydrogen-bonded structure of a protein, but it only is concerned with hydrogen bonds that are formed between the peptide-bonded N–H and the peptide-bonded carbonyl (C=O) (see the red arrows in Figure 4.11).

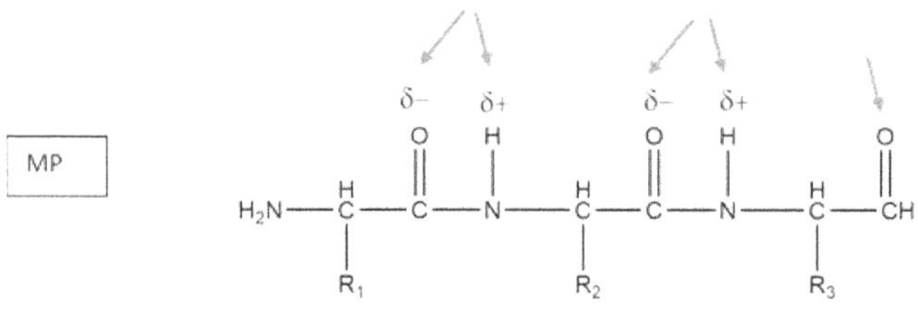

Figure 4.11. Secondary structure forms because it is stabilized by many hydrogen bonds. In both alpha helix and beta pleated sheets, every peptide-bonded N–H hydrogen and every peptide-bonded carbonyl oxygen (C=O) is part of a hydrogen bond (see arrows). The R groups are not involved in forming secondary structure.

Source: M. Poenie.

1) Alpha Helices

In secondary structure, adjacent N–H and C=O do not form bonds with each other. In the alpha helix hydrogen bonds form with the N-H hydrogen of one amino acid and the carbonyl oxygen of an amino that is four amino acids away on the chain. In the beta pleated sheet, hydrogen bonds form between two adjacent strands.

Alpha helices form a rod. The center of the helix is not really empty but filled by the atoms of the amino acids. Notice that the R groups point ouvtward from the center and so are out on the surface of the rod. This is important because the R groups of the helix can interact with other proteins.

a. Building Structures with Alpha Helices

Starting with the idea that an alpha helix is a rod, consider what happens if all the amino acids along one side of a rod are hydrophobic. This forms a sticky strip down one side of the rod. If you have a second helix that has the same arrangement of amino acids, then in water, the two hydrophobic strips will come together and stick. If, on the other hand, the sticky stripe of hydrophobic amino acids spirals around the alpha helix, when it encounters another similar alpha helix the two helices will twist around each other, creating a coiled coil.

An example of protein secondary structure: The a-helix

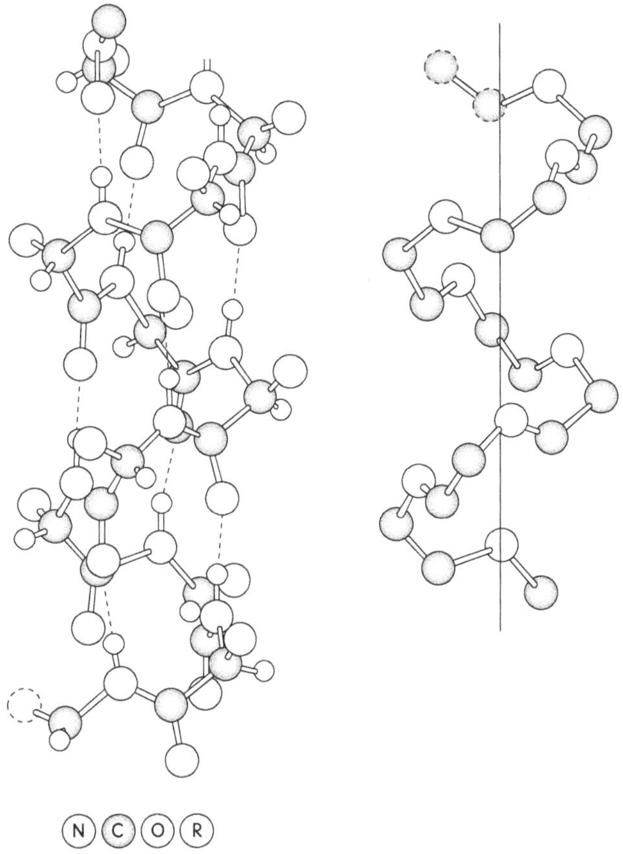

Figure 4.12. In the alpha helix, hydrogen bonds form between an N–H hydrogen and a carbonyl oxygen (C=O) four amino acids down the helix (see dashed lines). Note that in the alpha helix, all the R groups point outward.

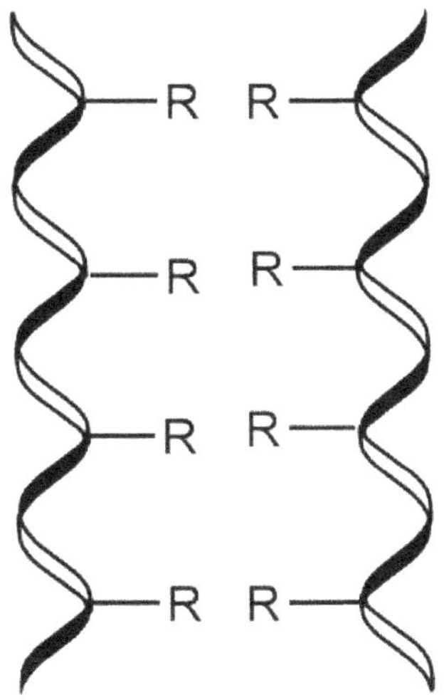

Figure 4.13. Building larger structures with proteins. Imagine that every time the amino acid R group came around to one side of the alpha helix, the amino acid R group was hydrophobic. This would generate a sticky stripe along the helix. In water, if there were another similar helix with hydrophobic R groups down one side, the two helices would want to stick together so that the hydrophobic R groups could both associate and get away from water.

Source: M. Poenie.

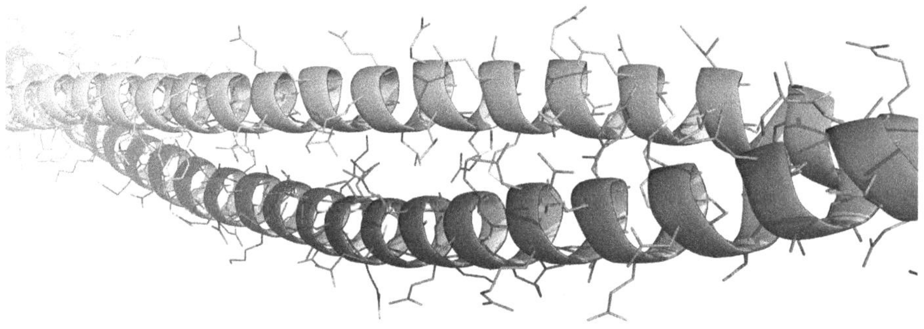

Figure 4.14. In second example, instead of having R groups straight down one side of the helix, imagine that there are two helices where the hydrophobic amino acids spiral around the alpha helix. In this case, in order to bring the hydrophobic amino acids together, the two helices have to wrap around each other, giving rise to a structure known as a coiled coil. The term coiled coil refers to two alpha helices, each representing a coil are wrapped around each other, to form a new coil.

2) Beta Pleated Sheet

The beta sheet is another type of *secondary structure* that also depends on every peptide bond carbonyl oxygen (C=O) forming a hydrogen bond with another peptide bond N–H hydrogen. In this case, instead of forming a helix, the amino acids are fully stretched out. The hydrogen bonds form side by side between the peptide strands. As these strands associate side by side, they give rise to a sheet. In the beta-pleated sheet, the R groups stick out above and below the sheet.

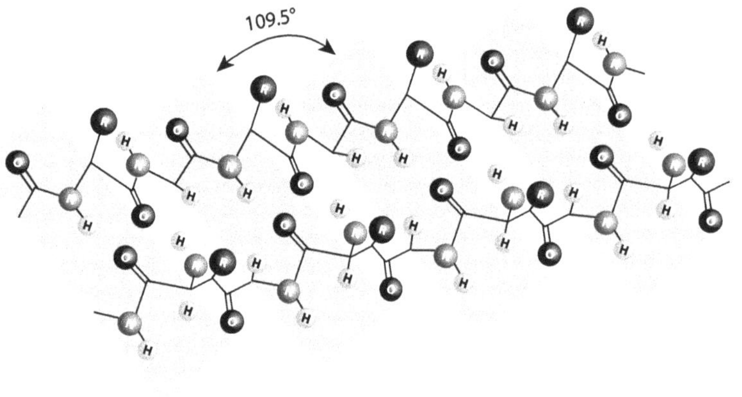

Figure 4.15. In the beta pleated sheet type of secondary structure, hydrogen bonds form side by side between the peptide chains. As peptide chains come together, the result is the formation of a sheet. The R groups are positioned either above or below the plane of the sheet.

3) Depiction of Secondary Structure in Proteins

It is common to use spirals to depict alpha helices and flat ribbons to depict beta strands with arrows to depict the structure of a protein (see Figure 4.16).

C. Tertiary Structure

Tertiary structure refers to the final folded state of the protein. Unlike secondary structure, which is not concerned with R group interactions, the tertiary structure *is*, due to R group interactions. R group interactions consist of weak bonds – hydrogen bonds, van der Waals bonds, and ionic bonds (sometimes called salt linkages). In addition, hydrophobic interactions drive hydrophobic amino acids away from the water and into the interior of a protein.

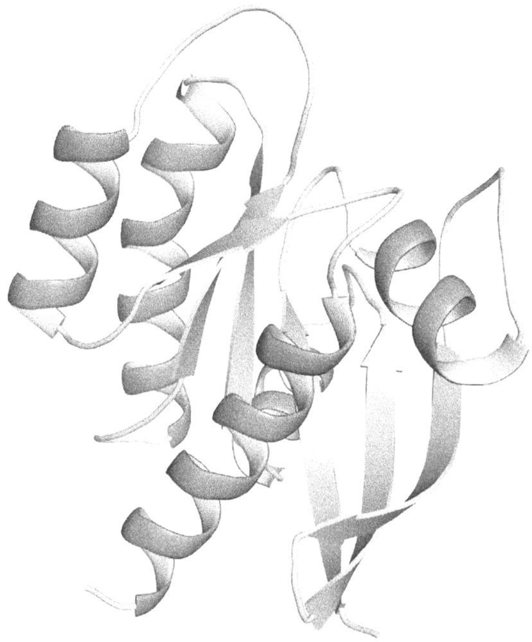

Figure 4.16. The general structure of a protein is often presented using coils to depict alpha helices (red) and flat ribbons to depict beta strands or beta pleated sheets (green).

D. Quaternary Structure

Quaternary structure describes the subunit arrangement when multiple proteins come together to make a functional whole. For example, if two copies of one type of protein come together with two copies of another protein to make the complete structure, the quaternary structure would be $\alpha_2\beta_2$, where alpha (α) and beta (β) refer to the two types of proteins.

E. Protein Domains

The concept of domains is critical to our modern concept of protein structure and must not be confused with quaternary structure. Domains concern the structure within one polypeptide chain. Up until now, we might think that protein folding involves the entire protein as one unit. However, regions of a polypeptide chain can fold into separate independent units which have different functions and/or different interactions. Regions of a polypeptide chain that fold independently from other regions of the same chain are called domains.

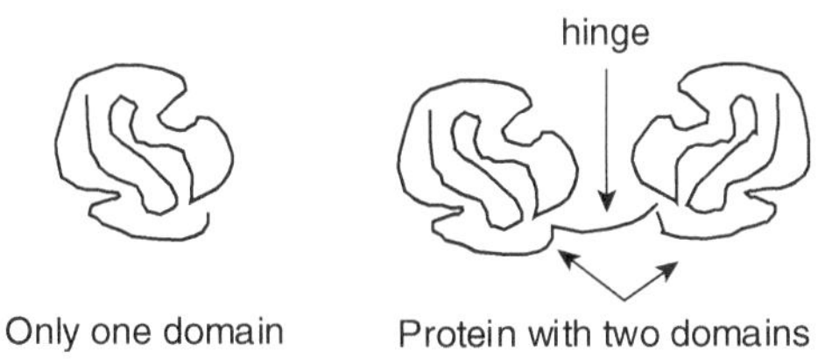

Figure 4.17.

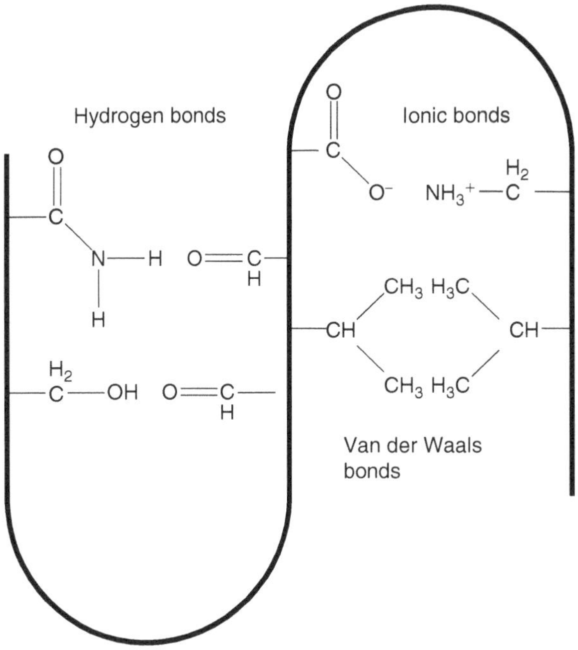

Figure 4.18. Can you identify the weak bonds holding the protein in its folded form?

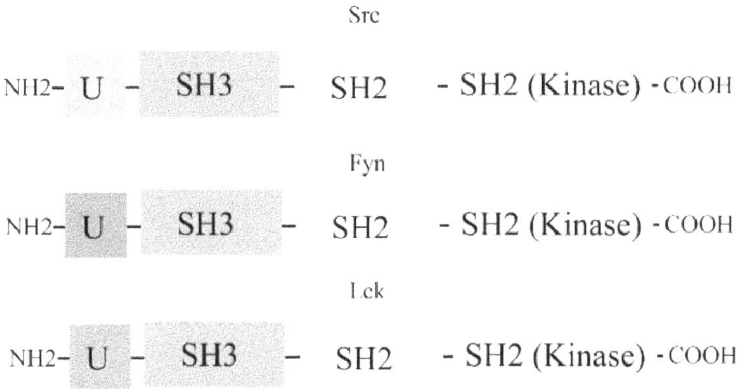

Figure 4.19. Domain structure of avian sarcoma tyrosine kinase (Src protein). Src is a tyrosine kinase composed of three domains known as Src homology (SH) domains. The kinase domain corresponds to SH1, a second domain called SH2 binds to tyrosine phosphates on proteins and a third domain (SH3) binds to polyproline regions of other proteins. We will discuss Src function in more detail later. There are a number of kinases that are very similar to Src and constitute the Src family of proteins. Each has the same domain structure and similar amino acid sequences. The main difference is at the N-terminus which is unique for each kinase.

Source: M. Poenie.

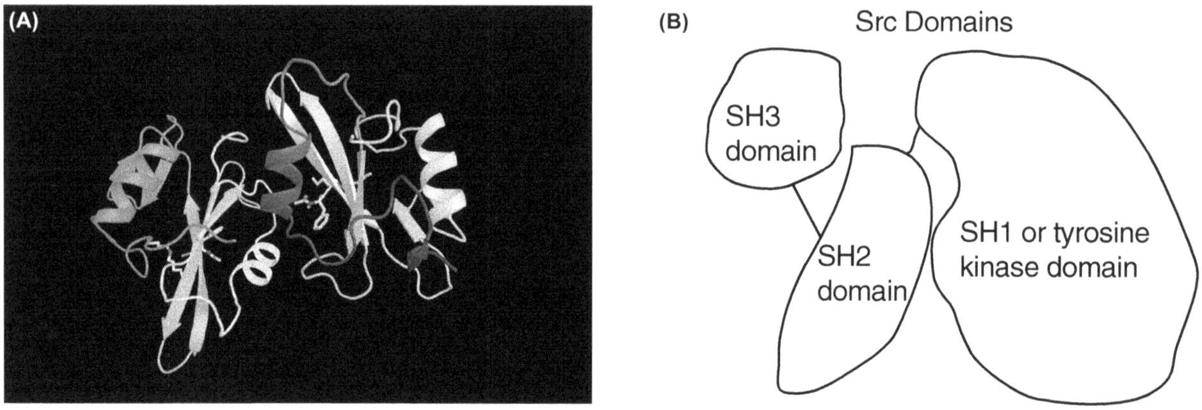

Figure 4.20. The Src protein is shown here using ribbons and coils to illustrate the structure. The kinase domain or SH1 is shown in blue. SH2 domain is shown in yellow and orange and SH3 domain is shown in red. Keep in mind that this structure involves only one peptide chain and each domain is connected to the next by a strand of the protein.

A: © ibreakstock/Shutterstock.com; B: © Kendall Hunt Publishing Company.

F. Domains and Exons

In eukaryotes the parts of a gene that actually code for the protein are often broken up into segments called exons. Between exons there are intervening sequences called introns. During transcription both the introns and exons are transcribed into RNA. Subsequently the introns are removed. Comparisons of the structure of genes and proteins reveal that in many cases one or two exons in a gene correspond to a domain of a protein. Furthermore, genetic recombination such as crossing over can lead to exchange of domains or transfer of domains if crossing over is unequal. This gives rise to the concept of exon shuffling.

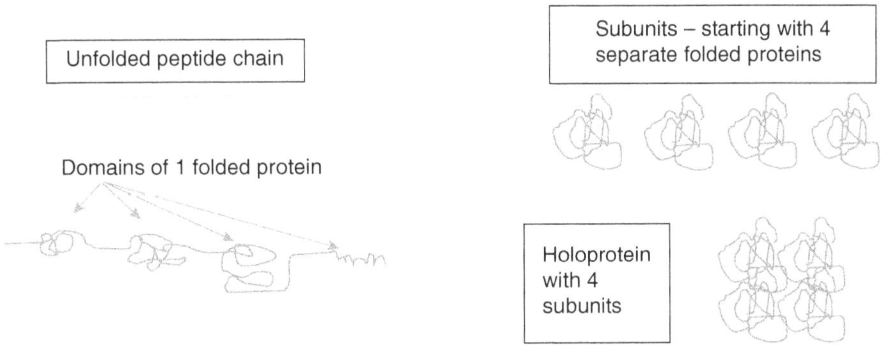

Figure 4.21. The difference between domains and subunits. Domains refer to regions within one polypeptide chain that fold independently of other parts of the peptide. Subunits refers to more than one polypeptide chain coming together to form the final protein (holoprotein).

3 Protein Folding

A. Proteins Must Fold to Function

The relationship between folding and function was revealed in classic studies by Christian Anfinsen. Anfinsen took the enzyme RNAase, which breaks down RNA, and treated it with urea and a reducing agent. Urea is a small molecule that makes numerous hydrogen bonds. At high enough concentration, urea competes with the normal hydrogen bonds formed between water and amino acids or between amino acids. Since the normal hydrogen bonds are integral to the structure of a protein, urea causes the protein to unfold. This is referred to as *denaturation* of the protein. When RNAase was denatured by urea, and its disulfide bonds reduced it lost enzymatic activity. Anfinsen then removed the urea by dialysis and found that RNAase activity returned to normal. Since all this can happen in a test tube, his results showed that *the sequence of amino acids contains the information needed for a protein to fold properly.*

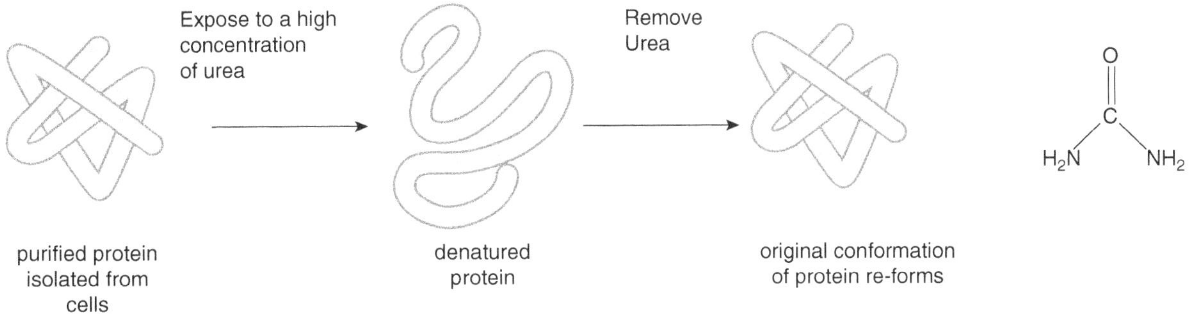

Figure 4.22.

B. Steps to Protein Folding

We now recognize a series of general steps proteins go through to reach the final folded state. The first step involves formation of secondary structure. Next there is the movement of *hydrophobic amino acid* residues to the interior of the protein. This is known as the *molten globule state*. Then, interactions between R groups give rise to the final folded form. The driving force for correct folding is achieving the most stable state (lowest free energy state).

C. Folding, Unfolding, and In Between

Proteins, especially those not stabilized by disulfide bonds, typically are easy to unfold. Unfolding or altering the normal folded state of a protein is called *denaturation*. Heating proteins usually denatures them, as seen when albumin of an egg turns white when heated. Physical turbulence, and chemical agents such as urea, alcohol, and detergents, can all denature proteins.

Although information is present in the sequence of amino acids that directs how a protein is to fold, this does not mean that newly made proteins always fold correctly. There are a huge number of theoretical folded states that a protein can adopt. Some of these may be relatively stable. This makes it possible for a protein to get stuck in an improperly folded state. Figure 4.23 illustrates the trajectory a protein might take en route to folding correctly.

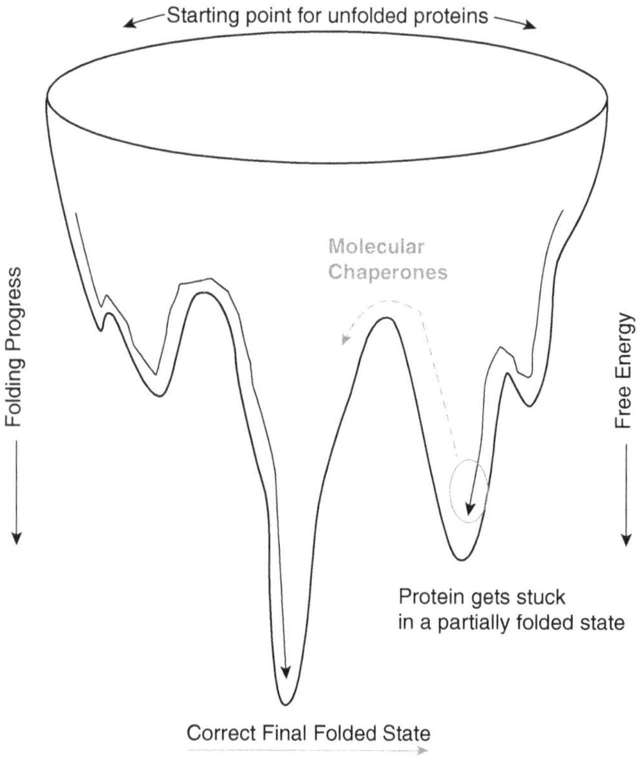

Figure 4.23. The trajectory a protein take to fold. A protein in a incompletely folded state can adopt many different partially folded states. This is represented by the top of the funnel. As folding proceeds, the protein moves down the funnel. However the path the protein takes may not be smooth and proteins can get stuck in a stable partially folded state as it progresses. Proteins stuck in these partially folded states may not be able to get out of them without the help of other proteins known as molecular chaperones.

D. Refolding Depends on Unfolding

Protein folding is energetically downhill (i.e., it has a negative ΔG). That does not mean that, starting out as a chain of amino acids, it will always get to the right folded state. Proteins in the process of folding can get stuck in states that are stable but are not in the proper targeted folded state Figure 4.23. Getting stuck in one of these states is like falling into a well or a valley. To keep going toward the right folded state, the protein has to climb out of the well or valley. To climb out of the well means that the protein has to unfold and go back to a previous state and then start again. Since folding is an energetically downhill reaction, unfolding a protein will require an input of energy, usually in the form of ATP

E. Molecular Chaperones and Chaperonins

Unfolded proteins represent a danger to the cell. This is seen in examples of prion disease as well as Parkinson's, Huntingdon's and Alzheimer's diseases. Each is an example of misfolded proteins causing cell death. When proteins are misfolded, they tend to expose hydrophobic regions to the exterior surface of the protein and this leads to aggregation of proteins. Once aggregates form, they can recruit more proteins to the aggregate. Ultimately, this can lead to cell death. In order to prevent this from happening, cells make a number of proteins that can bind misfolded proteins and prevent aggregation as well as helping misfolded proteins to fold correctly. These proteins are collectively known as molecular chaperones.

Molecular chaperones are a large family of proteins that aid in various aspects of protein folding. They can bind to newly synthesized proteins to prevent their folding before they are completely synthesized, they can bind to proteins being transported through membranes, they can bind and prevent aggregation, and they can help refold misfolded proteins. Many molecular chaperones act as indivdual proteins. However, a subset of molecular chaperones called chaperonins assemble into a large structure that serves as a folding cage. The chaperonins can be further subdivided into Group 1 chaperonins that are found in eubacteria, mitochondria and chloroplasts and Group II chaperonins which are found in archaea and the cytoplasm of eukaryotes.

Molecular chaperones were discovered in studies of heat shock. The heat shock response is a response to stresses such as hypoxia, toxins and elevated temperatures. It can be triggered by a sudden increase of even a few degrees C which is sufficient to cause some proteins in the cell to become misfolded. The small temperature change it takes to cause misfolding shows how delicate the folded state is. During the heat shock response, synthesis of a set of proteins called heat shock proteins (HSPs) is rapidly induced. Many of these heat shock proteins are designated by the name HSP and their molecular weight such as HSP70 and they are generally involved in dealing with misfolded proteins.

Molecular chaperones that act individually.

Two of the best characterized individual molecular chaperones are HSP70 and HSP90. HSP70 can act either on newly synthesized proteins or proteins that have later become unfolded. It works through ATP-dependent cycles of binding and unbinding to the hydrophobic core of an unfolded protein. As such, it can bind to largely unfolded proteins either preventing them from folding or through cycles of binding and release it can unfold proteins and then release them to allow for repeated chances at folding correctly. HSP70 works like a tag team with HSP90 where HSP90 acts at later stages when a protein is nearly folded. HSP90 also utilizes ATP for binding and unbinding but how it recognizes the unfolded protein is less clear.

Chaperonins

The best known of these is the bacterial GroEL and GroES complex of proteins. Chaperonins are part of a multiprotein complex that forms a cage that binds to unfolded proteins and helps them refold correctly. The Figure 4.24 illustrates the general process involved in the refolding of a misfolded protein. The protein binds to the folding cage probably due to exposed hydrophobic regions. The energy of ATP is then used to unfold the protein and release it. This allows the protein another chance to fold correctly.

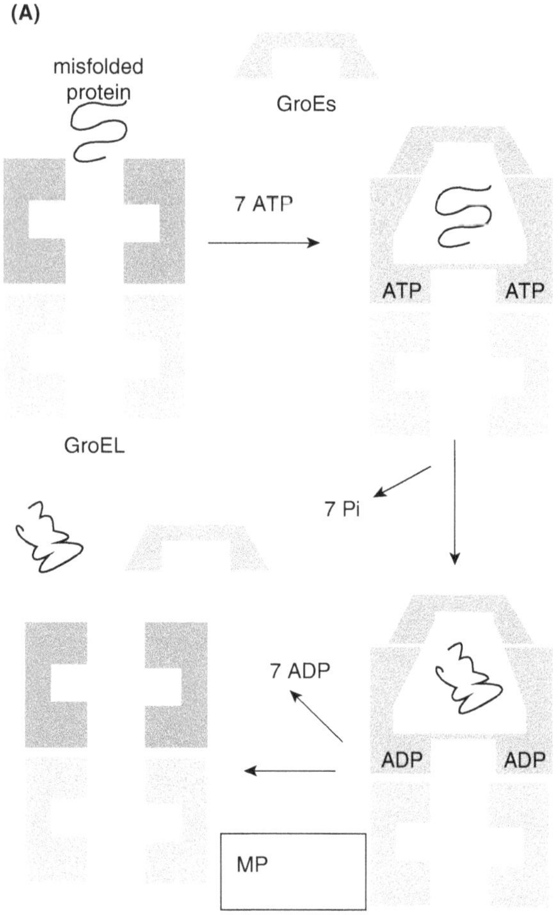

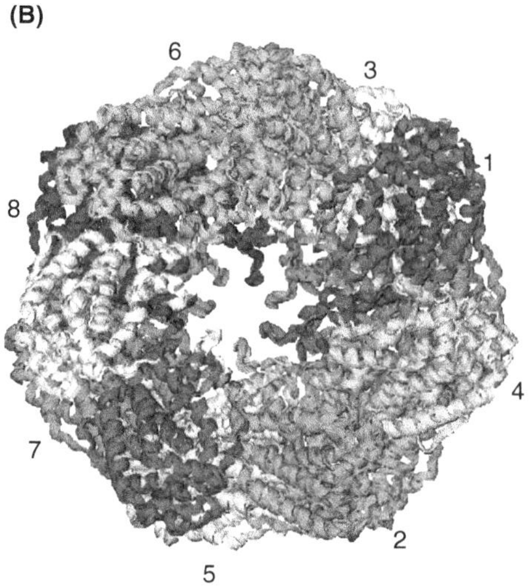

Figure 4.24. Protein folding by chaperonins. GroEl forms a chamber and GroES for a lid to the chamber that is used to refold misfolded proteins. While there is currently no consensus on the exact details of how protein refolding takes place, it is generally thought that proteins bind to the chamber due to exposed hydrophobic regions and the energy of ATP is used to unfold the protein. This gives the protein another chance to fold correctly.

A: © Kendall Hunt Publishing Company; B: Mark Lorch/ Shutterstock.com.

4 Ubiquitylation, or, What Do You Do if You Can't Fix It?

Cells, for all their complexity, like to keep their space neat and tidy. They do not tolerate junk lying around. When proteins don't fold properly. it becomes junk that, if the cell does not deal with it, can precipitate out of solution, and then it really becomes difficult to deal with. Furthermore, it can lead to death of the cell. The solution to this problem is to add a chain of small proteins known as ubiquitin.

There are systems in the cytoplasm that deal with misfolded proteins that cannot be successfully refolded. One of them involves ubiquitylation of proteins and their destruction in the proteasome. However, it should be noted that, addition of ubiquitin to a protein can mean a number of different things depending on how many ubiquitins are added and how they are attached. Our focus will be on formation of a ubiquitin chain that targets a protein to the proteasome.

A. Ubiquitylation

When the cell's complement of molecular chaperones and chaperonins cannot successfully refold a protein, that protein is tagged for destruction by ubiquitylation and then transfer of the ubiquitylated protein to the proteasome. Ubiquitylation is the process of attaching the small protein

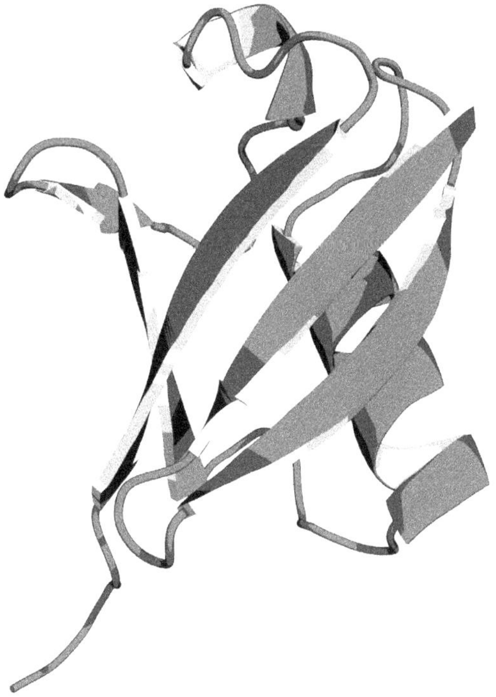

Figure 4.25. Ubiquitin is a small protein that can be attached to other proteins. To attach it, the carboxyl group is first activated by E1 enzymes. Ultimately, this activated carboxyl is attacked by the ε-amino group of a lysine on another protein, forming an attachment. Ubiquitin has seven lysines that can each be a point of attachment for other ubiquitins. The result is that a wide variety of different ubiquitin chains can be formed.

ubiquitin to the damaged protein. The ubiquitin protein is attached to lysines on the target protein and often the process goes through several cycles to produce a chain of ubiquitin molecules.

1) The Ubiquitin Code

The basic process of attaching ubiquitin to another protein is carried out by a series of three ubiquitylating enzymes called E1, E2, and E3. In humans there are two E1 enzymes, forty E2 enyzmes, and six hundred E3 enzymes. The large number of E3 enzymes reflects the diverse targets and ways ubiquitylation designates function. Ultimately, the chain of ubiquitin molecules binds to a cytoplasmic structure called the proteasome that destroys the protein.

2) Ubiquitylation by E1, E2, and E3 Ligases

Ubiquitylation is accomplished through a catalytic cascade involving ubiquitin-activating enzymes (E1), ubiquitin-conjugating enzymes (E2), and ubiquitin ligases (E3). The human proteome contains two E1 enzymes, approximately forty E2 enzymes, and over six hundred E3 enzymes, the combination of which accounts for the large variety of ubiquitin modifications. The large number of E3 enzymes makes sense when we realize that there are thousands of different proteins that can be ubiquitinated.

a. Ubitquitin Ligase E1

Before ubiquitin can be attached to a target protein it first needs to be activated, since the ubiquitin carboxyl group that is attached to Lys48 on the target protein is chemically inert. To activate the carboxy group on ubiquitin, E1 first causes a reaction with ATP. The carboxyl group attacks the first (or alpha) phosphate on ATP and displaces the first two phosphates. Note that displacing two phosphates instead of the usual one phosphate is a way of generating more energy from ATP. The remaining adenosine monophosphate (AMP) is now connected to the ubiquitin carboxyl group. The AMP is then replaced by a cysteine SH group on E1 to give a thioester linkage, which is still considered a high-energy linkage.

b. Ubiquitin Ligase E2

Ubiquitin Ligase E2 receives the activated ubiquitin from E1. The transfer is from one SH group to another so the ubiquitin remains activated. Once the activated ubiquitin is attached to E2 it is either transferred directly to target proteins or it is transferred to E3.

c. Ubiquitin Ligase E3 Enzymes

What happens now depends on which E3 enzymes are involved – either the HECT family or RING family. There are over six hundred E3 enzymes, most of them belonging to the RING family. The large number of E3 ligases reflects the diversity of proteins that they ubiquitinate and the different types of cellular activities they regulate. The RING family and Hecht families transfer ubiquitin to a target protein in different ways. In the case of RING E3 proteins, E3 acts as a scaffold that brings E2, E3, and the acceptor protein together in order to catalyze the transfer of ubiquitin from E2 to the target protein. For the HECT family of E3 ligases, the ubiquitin is first transferred from a cysteine on the E2 to a corresponding cysteine on E3. It is then transferred from E3 to the target protein.

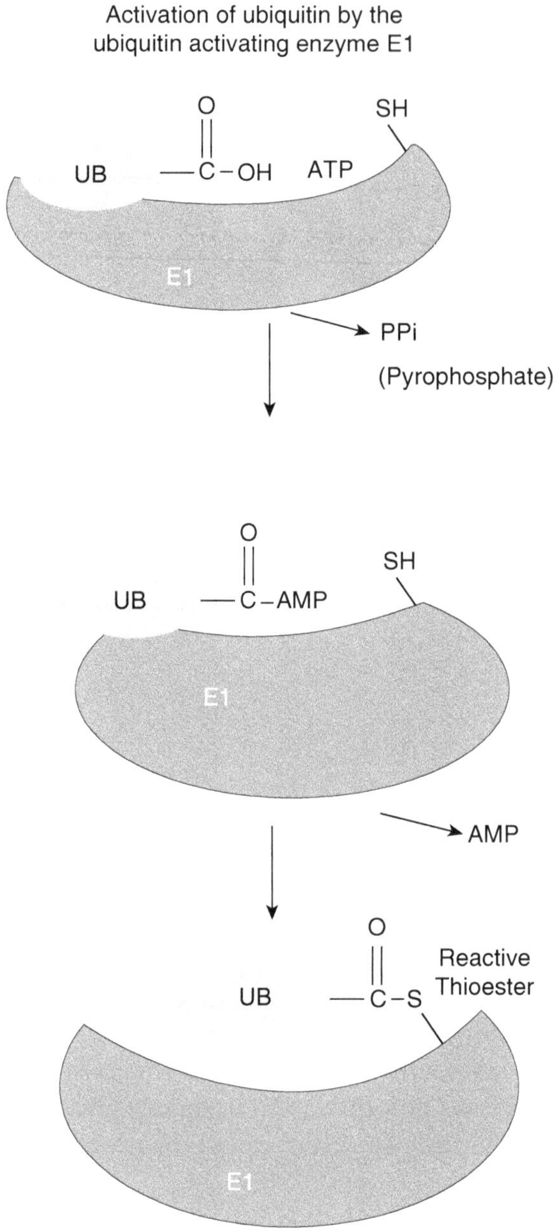

Figure 4.26. The ubiquitin activating enzyme E1 uses ATP to make a relatively inert carboxyl group into a relatively reactive thioester. It does this by first reacting the carboxyl with ATP to give ubiquitin attached to adenosine monophosphate (AMP). From here the ubiquitin is transferred to an SH group on the E1 enzyme to give a thioester. This thioester is much more reactive than the original carboxyl group such that ubiquitin can now be transferred to other proteins.

B. The Proteasome

The proteasome is a structure that contains two caps (shown in purple in Figure 4.29) and a central chamber (core shown in yellow) whose walls contain a number of different proteases. The cap recognizes the ubiquitin tag, binds, and proceeds to unfold the protein and drive it into the core using the energy in ATP. During this process, the ubiquitin is cut off from the protein. The core then digests the protein into its component amino acids.

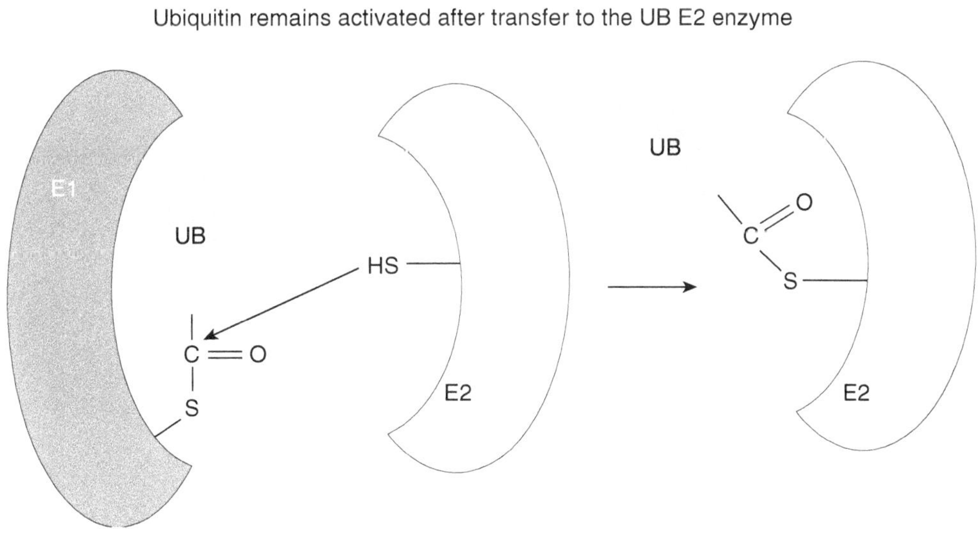

Figure 4.27.

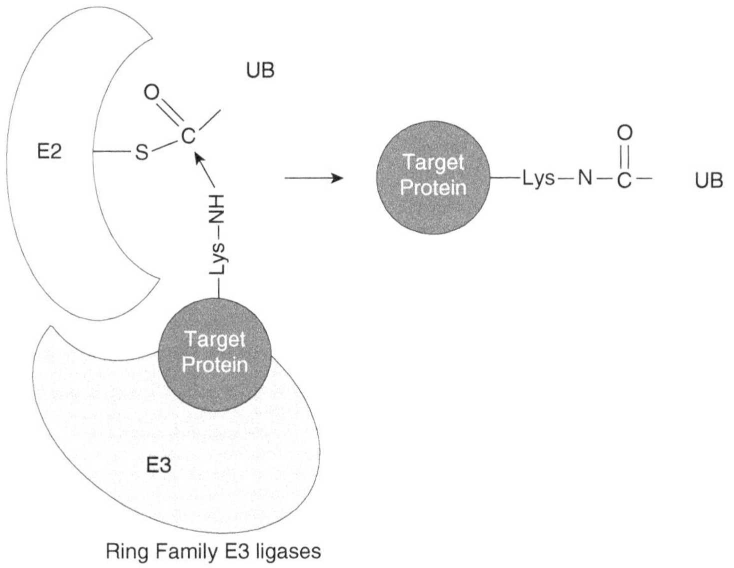

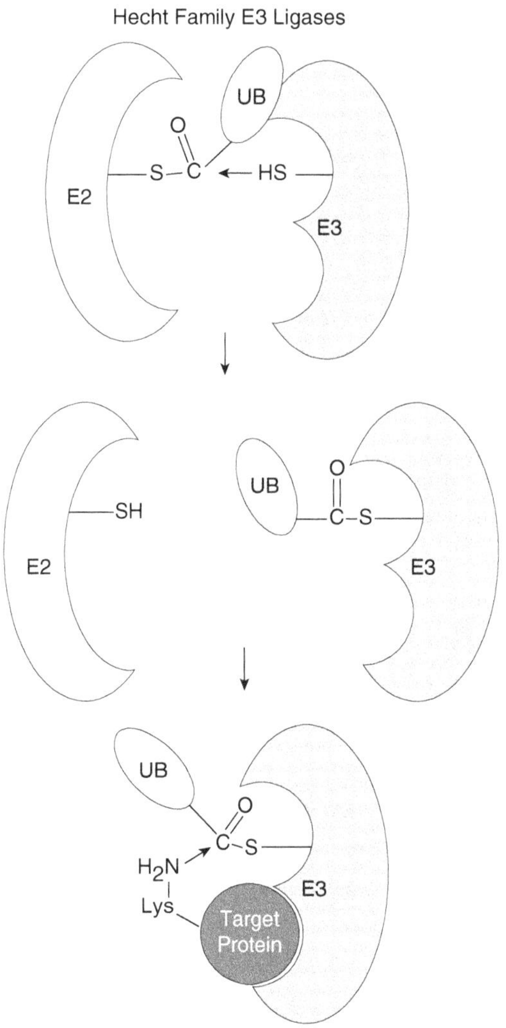

Figure 4.28. E3 ligases. There are two families of E3 ligases, known as the RING family and the Hecht family, which operate differently. In the RING family the E3 serves to catalyze the transfer of the ubiquitin from E2 to the target protein. In the Hecht family the ubiquitin is first transferred from E2 to the E3 and then to the target protein.

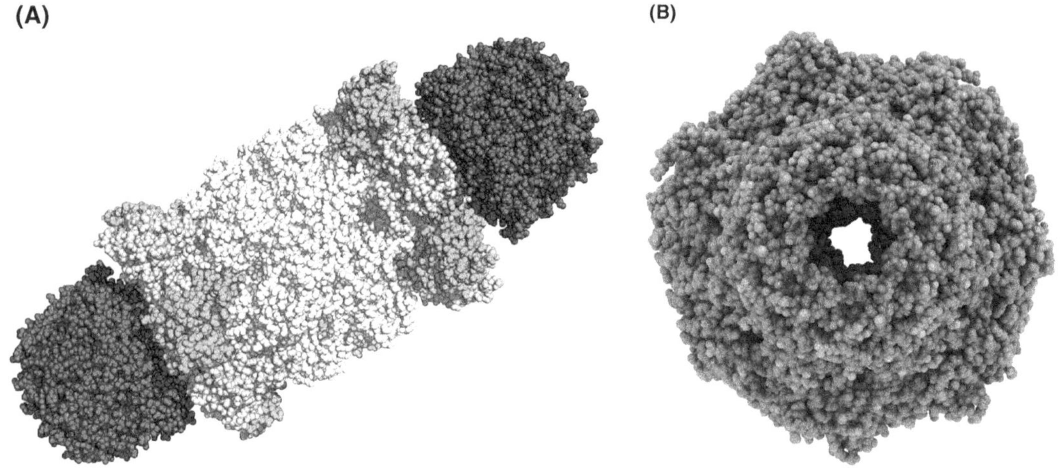

Figure 4.29.

A: © petarg/Shutterstock.com; B: © molekuul_be/Shutterstock.com.

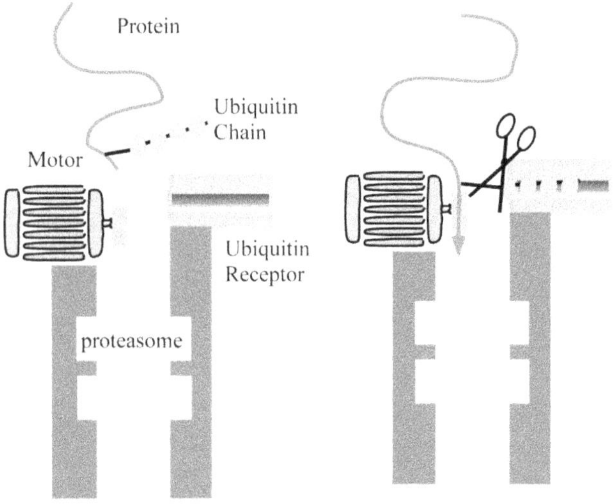

Figure 4.30. The proteasome is a chamber lined with proteases that chop up proteins as they go through it. Proteins are targeted to the proteasome through an attached chain of ubiquitin molecules. Once the ubiquitin is recognize a motor complex draws the protein into the proteasome and the ubiquitin chain is detached.

Source: M. Poenie.

5 Methods for Protein Analysis

A. SDS Polyacrylamide Gel Electrophoresis (SDS PAGE)

SDS PAGE is a procedure that separates proteins according to their molecular weight. It depends on the use of a detergent sodium dodecylsulfate and a reducing agent (beta-mercaptoethanol or dithiothreitol) that reduces and breaks up disulfide bonds. The sulfate on the SDS molecule has a fixed negative charge. The number of SDS molecules that bind to the protein backbone are in constant proportion to the size or molecular weight of the protein. As a result, the amount of negative charges from the detergent relative to the mass of the protein is a fixed proportion. In other words, the charge-to-mass ratio is the same for all proteins, whether big or small. However, different size proteins still differ by their mass. The difference in masses is the remaining variable and the basis for protein separation on an SDS PAGE gel.

Proteins bound to SDS have many negative charges. When a negative electrode is placed at the top of a gel and a positive electrode is placed at the bottom, proteins will migrate through the gel. The gel itself is a mesh made of acrylamide that proteins have to wiggle through. The more concentrated the gel, the harder it is for proteins to move through, such that smaller proteins will normally always move through the gel matrix faster than larger proteins.

Electrophoretic gels are normally composed of two different gels, a larger lower gel called the separating gel and an upper gel called the stacking gel. The composition and pH of the buffer in the stacking gel and the tris-glycine in the running buffer cause proteins in the stacking gel to all squeeze together and move in unison as a very thin band. The advantage of this is that when proteins move into the separating gel, they start out as very thin bands. As proteins move through the separating gel, these bands will spread or fatten. However, the thinner the protein bands are, the easier it is to resolve them from each other. Overall, then, by using the combination of of a stacking and separating gel, we get better resolution of the proteins on the gel.

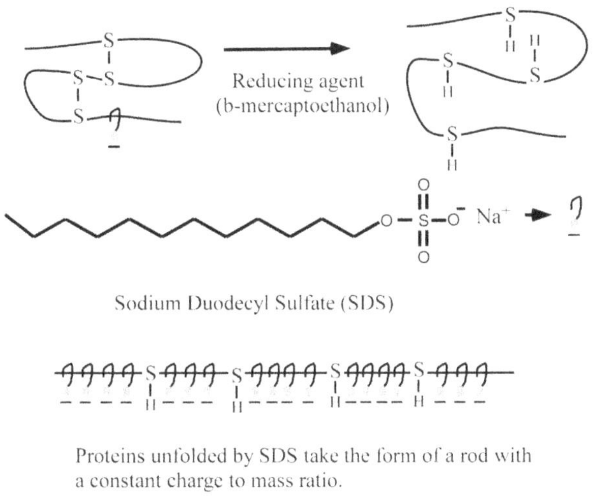

Figure 4.31. Protein reduction and denaturation by the sample buffer. In order to run properly, proteins need to be reduced to remove disulfide bonds. Proteins are unfolded by SDS molecules that wrap around the protein backbone. The number of SDS molecules that bind is proportional to the size of the protein, resulting in a constant charge-to-mass ratio.

Source: M. Poenie.

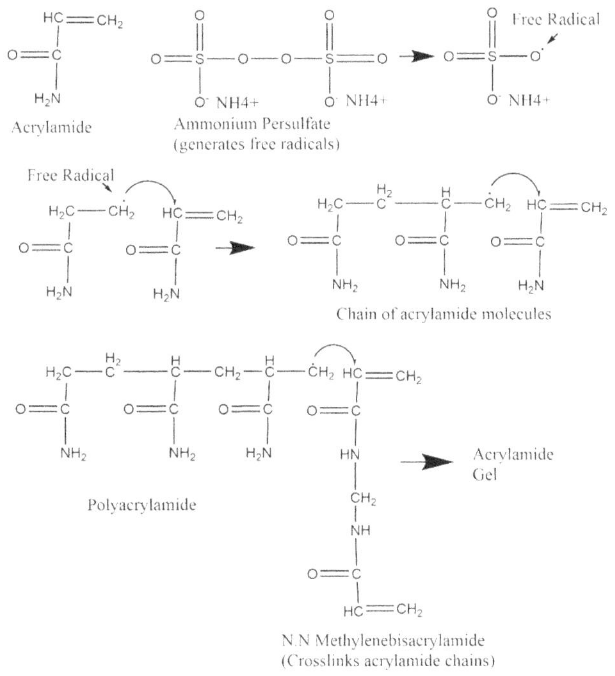

Figure 4.32. Formation of polyacrylamide gels. Acrylamide gels are formed by the free radical polymerization of acrylamide monomers. Free radicals are generated first by the breakdown of ammonium persulfate. The ammonium persulfate abstracts an electron from acrylamide, which then attacks a second acrylamide monomer. The addition of a small amount of bisacrylamide results in the formation of a three-dimensional gel matrix.

Source: M. Poenie.

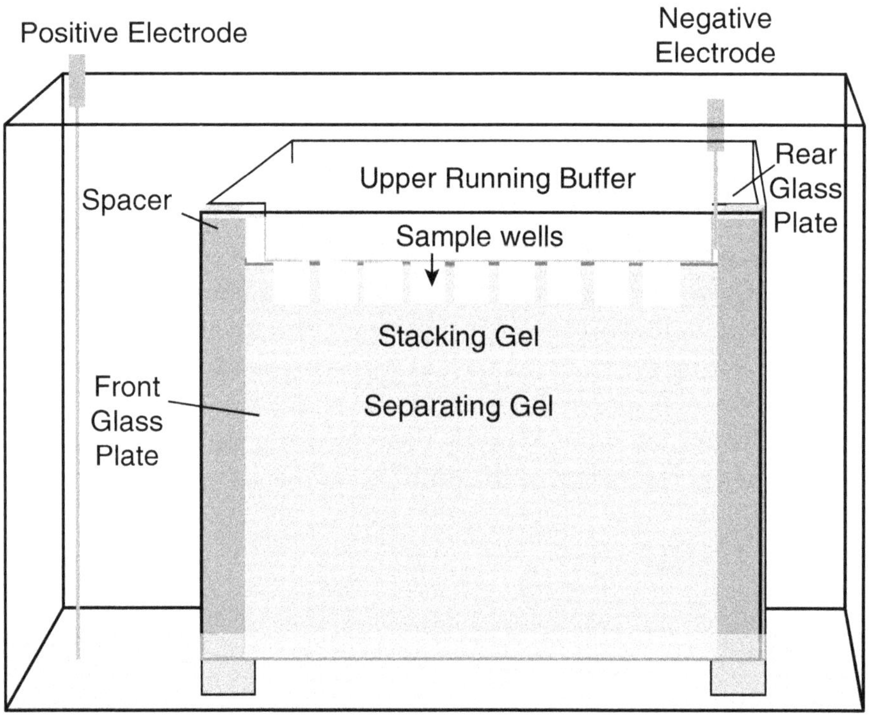

Figure 4.33. A polyacrylamide gel electrophoresis apparatus. In polyacrylamide gel electrophoresis, an acrylamide solution is poured between two glass plates separate by spaces. The gel is poured in two stages, first the lower gel known as the separating gel. Once the gel has polymerized a second solution forming the stacking gel is added on top the separating gel. The purpose of the stacking gel is to force all the protein into a very thing band before they enter the separating gel. This provides for better resolution of the proteins because the bands are thin. Without the stacking gel, protein bands could be quite thick. The glass plate and gel assembly is then placed in the gel apparatus so that both the upper and lower surfaces of the gel are exposed to separate chambers which each contain the solution called the running buffer. Protein mixtures are added into the sample wells. The electrodes are attached to a power supply such that the upper chamber receives the negative electrode and the lower chamber has the positive electron. When power is applied, proteins move from the top to the bottom and begin separating when they enter the separating gel. Since the buffers and the protein mixture usually contains SDS, proteins will migrate according to their molecular weight with the smallest proteins moving the furthest.

B. Western Blot (immunoblot)

The term "Western blot" arose from the fact that two other blotting techniques were called Southern (the name of the inventor) and Northern (as opposed to Southern). Northern and Western blots are nucleic acid blotting techniques, whereas the Western blot is for proteins. In a Western blot, one starts out with an SDS PAGE gel. After running the SDS PAGE gel, the proteins in the gel are transferred to a nitrocellulose membrane Figure 4.4. When proteins are in the gel, they are not accessible to other proteins. Transferring them to a nitrocellulose membrane puts the proteins on the surface of the membrane, where they are accessible.

To transfer proteins over to the membrane, a sandwich is made where the gel is placed on the surface of the nitrocellulose sheet and the two are held together in a cassette sandwiched between a mesh material that allows free passage of solution and electrical current. Then power is applied to the sides of the cassette, which causes the proteins to move out of the gel and over to the nitrocellulose sheet, which binds to the proteins. The idea is to move the proteins quickly over to the nitrocellulose sheet without having them spread out or diffuse away. Ideally you will have the proteins showing the same banding pattern on the nitrocellulose sheet as you had in the gel.

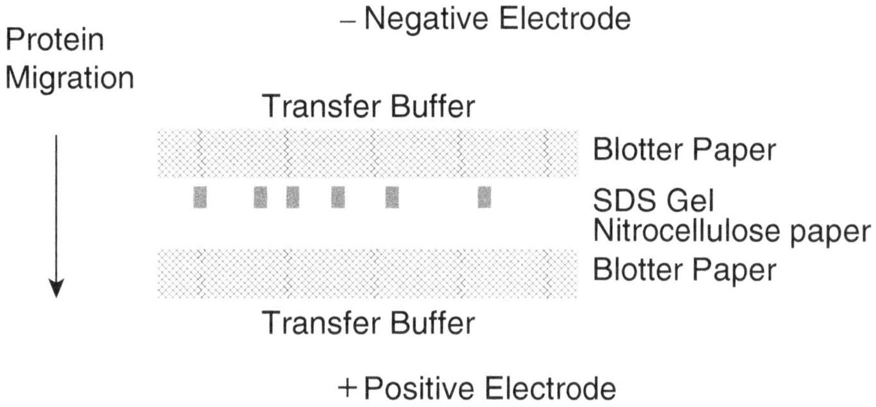

Figure 4.34. Western blotting In a Western blot, proteins are separated by polyacrylamide gel electrophoresis as described above. The polyacrylamide gel is then laid on a sheet of nitrocellulose paper and sandwiched between various types of mesh material such as blotter paper. This sandwich is then placed in a apparatus that allows for electrical current to pass through the sandwich. The current drives the proteins out of the gel and onto the nitrocellulose paper where they become firmly attached. Once the proteins are out of the gel and on the surface of the nitrocellulose paper, they are exposed and available for further processing.

Visualization of Proteins on the Blot

Once proteins have been transferred to the nitrocellulose sheet, one can add primary antibodies to a protein of interest. If the protein of interest is on the nitrocellulose sheet, the antibodies will bind to the band. One can then follow with a secondary antibody that binds to the primary. This antibody (usually) is conjugated to the enzyme horseradish peroxidase (HRP). When horseradish peroxidase reacts with certain compounds, it can either give off light or generate a colored product. Most popular today is to use the light-emitting reaction. Here the membrane is soaked in a solution containing luminol and then the membrane is sandwiched to a piece of X-ray film and placed in the dark in a light-tight container. Wherever there is a band that has HRP, light will be given off and will expose the X-ray film. The film is then developed to show the bands that are present.

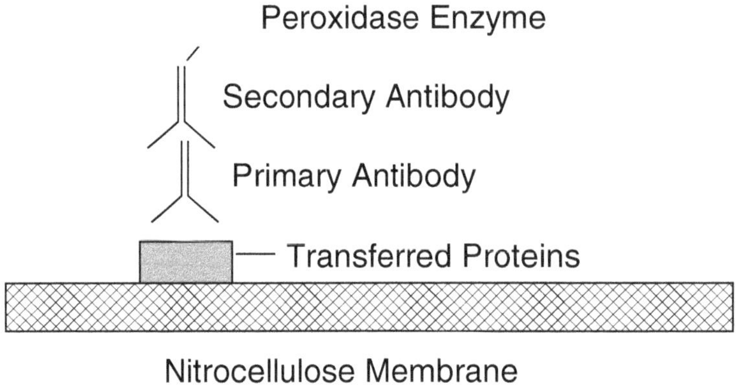

Figure 4.35. Visualization of proteins on nitrocellulose using antibodies. The nitrocellulose sheet is incubated with antibodies to a particular protein (primary antibody), washed and then incubated with a second antibody that binds to the first. Horseradish peroxidase is attached to the second antibody and this enzyme catalyzes a chemical reaction that lets one see the protein bands.

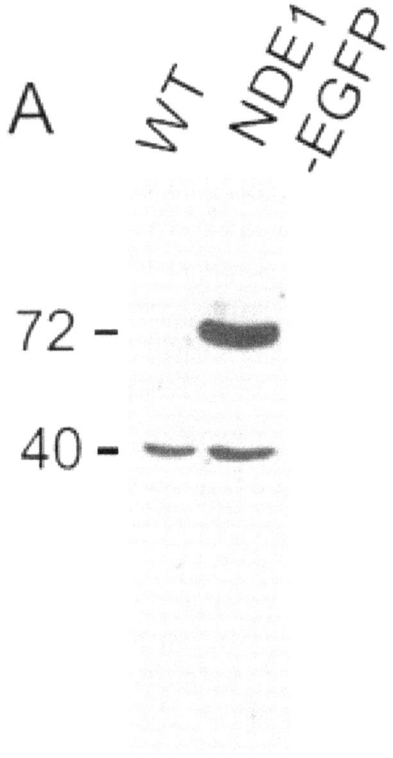

Figure 4.36. A finished Western blot. This figures shows a strip of nitrocellulose that has been exposed to antibodies as described in 4.35 above and developed to show where the antibodies bound. In this particular example, the nitrocellulose membrane was probed with an antibody against NDE1. On the left proteins were taken from wild-type cells that express normal NDE1 (40 Kd). On the right, proteins were taken from cells that express both normal NDE1 and NDE1 fused to green fluorescent protein (72 Kd).If this blot were stained to show all proteins present in the original SDS PAGE gel, there would be hundreds of bands.

Source: M. Poenie.

Enzymes

5

1 Catalysis

A. Chemical Reactions

Eukaryotic cells carry out roughly a billion chemical reactions per second. These reactions keep the cell going and supplied with all the things it needs to stay alive and grow. Each chemical reaction involves an overall decrease in free energy (i.e., it has a negative ΔG). However, even though a chemical reaction has a negative ΔG, that does not mean that it will proceed quickly. We need to distinguish between the thermodynamics of a chemical reaction and the kinetics of the reaction. The thermodynamics of a chemical reaction concerns the equilibrium concentrations of a reaction given the nature of particular reactants. For our purposes it is expressed in terms of the free energy change of the reaction. Reactions with a negative free energy go forward as written, whereas reactions with a positive free energy go backward. When a reaction is at equilibrium, the free energy is zero. However, that a given reaction has a negative free energy does not tell you anything about how fast that reaction will go.

The rate of a reaction is dependent on several factors. These include the concentration of the reactants and the temperature, pressure, and activation energy for the reaction. However, in the cell, the most relevant is the activation energy. Many molecules in the cell, including proteins, lipids, chains of carbohydrates, and nucleic acids, all have a negative free energy of hydrolysis. So we know these molecules will eventually end up being hydrolyzed into their component monomers. But that does not mean that they will break down quickly. We call those type of molecules "metastable." They are thermodynamically unstable but they are kinetically stable. The reason for this is that there is a large activation energy for hydrolysis.

1) Energy of Activation

Enzymes are proteins that catalyze or speed up reactions. They do this by lowering the energy of activation, E_A, *the energy that must be put into a molecule or system to get the reaction started.* In a chemical reaction molecules have to collide to react, and these collisions must be sufficiently energetic to get the reaction started. We might look at this from the vantage point of asking what fraction of the reactants have enough kinetic energy to overcome the activation barrier, and how they do it. If a large percentage of the molecules have enough energy to react, the reaction will go quickly. If few molecules have enough energy to react, then the reaction will go slowly. However, if we lower the activation energy so that a greater fraction of the molecules have enough energy to react, the reaction will go faster. This is what enzymes do.

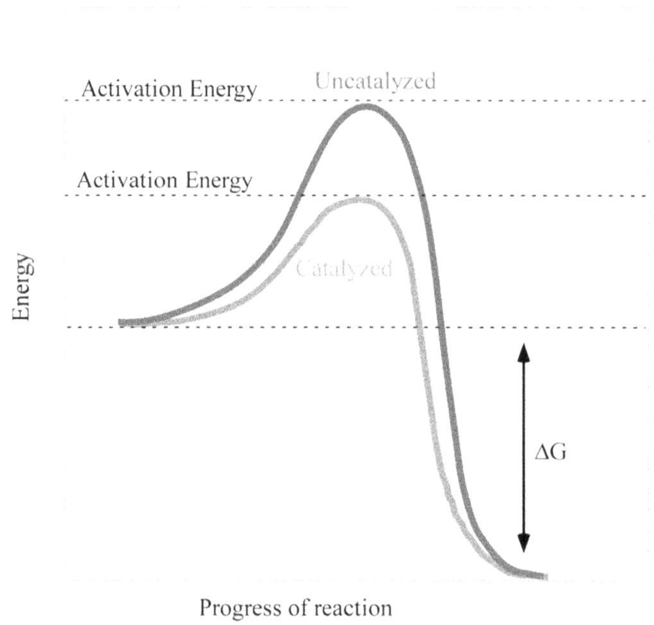

Figure 5.1. Plot of the free energy of reactant molecules during a reaction. In order for a chemical reaction to occur, molecules must overcome the activation energy barrier. This means that molecules must collide in the proper orientation and with enough force for the reaction to take place. The rate of the reaction is determined by the number of molecules in the system that have enough free energy to react. To increase the rate of an uncatalyzed reaction would commonly involve heating to increase the rate of collisions. The presence of a catalyst lowers the activation energy so that a greater number of molecules have enough free energy to react.

Source: M. Poenie.

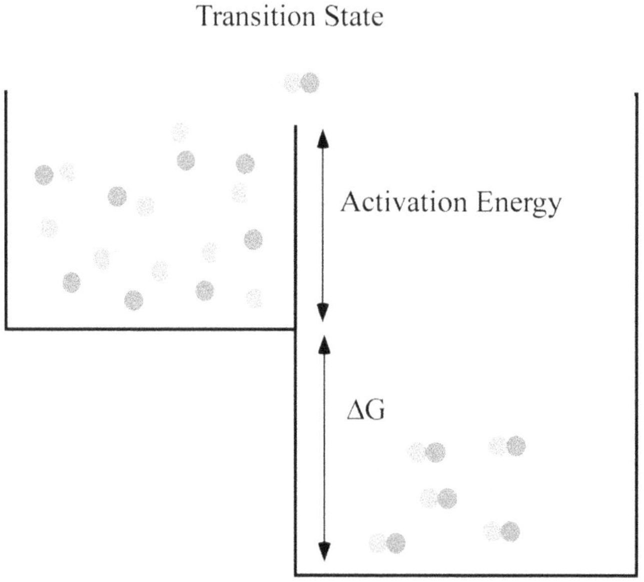

Figure 5.2. Chemical reactions involve the collision of molecules. These collisions must be sufficiently energetic to get the reaction started. The required energy is represented as the activation energy barrier. For the reaction to go forward, molecules must get over this barrier.

Source: M. Poenie.

B. How Do Enzymes Speed Up a Reaction?

There are a number of things enzymes do that contribute to lowering the activation energy.

1) The Cage Effect

The cage effect refers to bringing the reactants together with the right orientation. In order to react, there must be a collision of reactants with sufficient energy to cause the reaction. In the absence of an enzyme, reactants would be moving around in solution and occasionally colliding. The more concentrated the reactants in solution, the more likely it is that they will collide. In binding to the substrate/reactants, they are held close together.

2) Strain

We will discuss this effect in more detail in Section C, but it is a general and significant component of how enzymes work. Enzymes do not simply bind to the substrate. Enzymes bind best to the transition state. So initially enzymes bind more weakly, but in binding both the enzyme and the substrate change shape. The enzyme binding progressively forces the substrate to change its shape into what looks like the transition state and as the substrate changes, the binding to the enzyme gets tighter and tighter.

3) Charge Effects

This aspect covers a number of effects. Enzymes can do acid/base catalysis. This means putting an H+ or an OH− in position to attack a bond or to cause electrons to shift. Acid/base catalysis can occur in solution, but it must be either acid or base catalysis. What is remarkable about enzymes is that they can do both acid and base catalysis at the same time. Enzymes can also use metal ions to pull electrons, making bonds more susceptible to breaking or attack.

C. The Transition State

This discussion relates to the idea of strain mentioned above. In a chemical reaction we typically break some bonds and form some new chemical bonds. We might illustrate this using a common type of chemical reaction where one atom displaces another. Our starting molecule might be a carbon atom with four different groups attached to it with (see Figure 5.3A). Coming from the right is a purple atom or group that is attacking the carbon molecule (M). As the reaction proceeds the purple atom comes closer and begins to form a bond with the carbon atom. Meanwhile the bond between carbon and another group (gray) begins to break. Eventually the gray atom leaves, being replaced by the purple atom (D). Somewhere in between A and C is the intermediate state, B, where the bond to the purple atom is partially formed and the bond to the gray atom is partially broken. We will call this the transition state.

For the moment, we can think of the transition state as in the middle, between the reactants (A) and products (C). Furthermore, for the moment, we will assume that the transition state is at the peak of the activation energy barrier. Once we get to this point, the bound reactants can fall apart back to free reactants or they can fall forward to products. The transition state represents that tipping point.

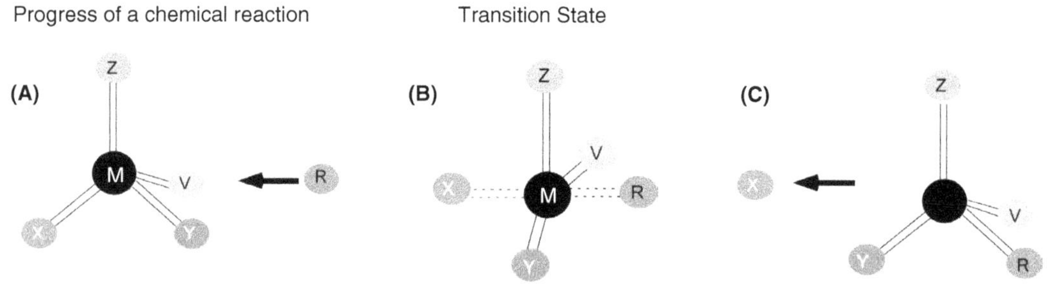

Figure 5.3. Progress of a chemical reaction. **(A)** In this illustration of a chemical reaction we start with a molecule composed of atom M with four substituent groups (X, Y, Z, and V). During the reaction, another atom or molecule (R) attacks and attempts to displace X from M. **(B)** As the reaction proceeds there is a point where a bond has partially formed between R and M and at the same time, a bond is breaking between X and M. This is what we call the transition state. Note that it is different from either reactant or product in that during the transition state there are five bonds to M, whereas as normally there are only four. **(C)** Eventually the bond between M and X breaks and the bond between M and R is fully formed.

Source: M. Poenie.

So what we want to say now is that enzymes bind to the transition state better than they bind to reactants or products. In Figure 5.4, there are three cases shown: no enzyme (A), an enzyme that has a binding site complementary to the substrate (B), and an enzyme that binds to something that is intermediate between reactant and product (the transition state) (C). Note that only state C lowers the activation energy.

D. Enzymes Bind to the Transition State

Enzymes must bind to a substrate in order to catalyze a chemical reaction. When an enzyme binds to a substrate, there is a change in the overall structure of the enzyme as it binds. The enzyme in turn, through binding, "squeezes" the substrate into the transition state. As the substrates adopt a shape like the transition state, the enzymes bind better and better. Once the enzyme has squeezed its substrate into the transition state, its job is essentially done. From this transition state, we can either go back to reactants, or the reaction can go forward to products. From the enzyme's point of view, it does not matter. It has done its job. What drives the reaction forward is the fact that the products have a lower free energy and energetically cannot come back and form reactants. This is illustrated in the figure below. Here we have an enzyme whose job it is to catalyze the breaking of a stick (stickase). So it is a crazy illustration, but now we have to imagine that there are atoms on the enzyme that are attracted to atoms on the stick. These points of attraction are weak chemical bonds that are forming. In Part B of the drawing, we consider what would happen if the enzyme were simply complementary to the stick. Here the enzyme would bind, but binding would not do anything to help the reaction. In fact, it would make the reaction more difficult, because the stick needs to bend before it breaks and binding of the enzyme would make it harder to bend. In Part C we have a more accurate depiction. In binding to the enzyme, the substrate is forced to bend. This bending brings the substrate closer to the transition state between being one stick and two sticks.

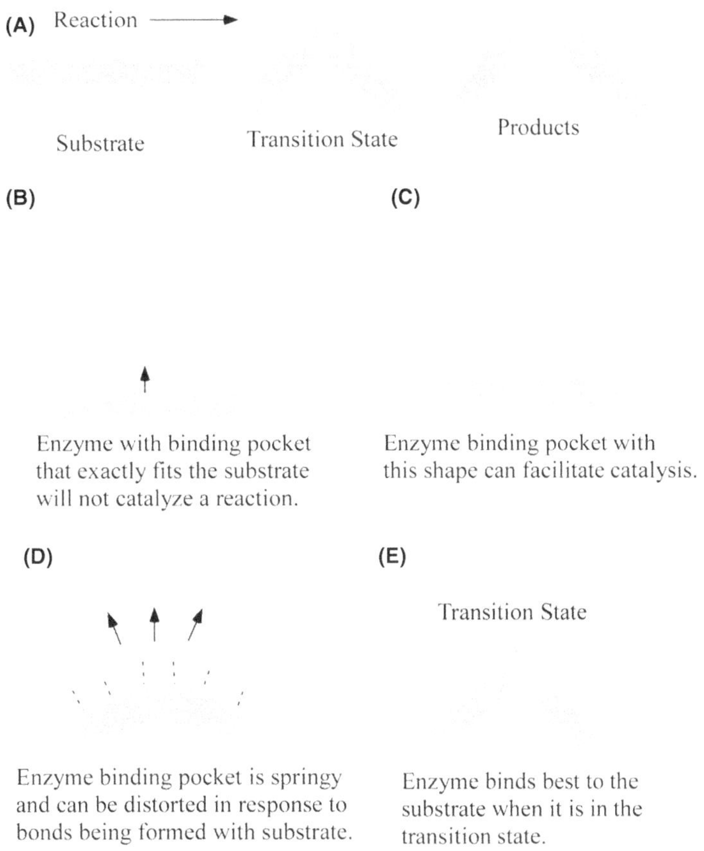

Figure 5.4. Model for enzyme catalysis. **(A)** Here we imagine the substrate as a rod that gets broken into two products. Between the substrate and the products is a transition state where the rod is bent but not broken. **(B)** An enzyme that binds perfectly to the substrate would actually inhibit the progress of the chemical reaction because here the enzyme wants to hold the substrate in the form it is in. It would take energy to distort the enzyme and the substrate into the transition state. **(C)** An enzyme whose optimal state binds to the transition state helps the substrate achieve the transition state. **(D and E)** Initial binding to the substrate is weak but as binding continues the substrate is warped into the transition state.

Source: M. Poenie. Adapted from Lenninger Principles of Biochemistry 6th edition WH Freeman.

1) Key Points

We want to summarize what we just went through with the following statements:

A. Enzymes lower the energy of activation by promoting the formation of the transition state.

B. Enzymes lower the energy of activation by binding more tightly to the transition state than to reactants or products.

2 Testing to See If Our Model of Enzyme Catalysis Is Correct

If enzymes catalyze reactions according to the schema we just presented, then there are implications, things we can test to see if our schema correct.

A. Test 1: Stable Transition State Analogs

The real transition state of a chemical reaction is very short-lived because it is very unstable. However, ingenious chemists have devised stable or nonreacting molecules that mimic the transition state of particular chemical reactions. These are known as stable transition state analogs. If the theory of enzyme catalysis is correct, these analogs should bind to the enzyme much more tightly than the normal substrate and in doing so, they should be potent inhibitors of the enzyme. This prediction turns out to be correct. In one example, in an attempt to counteract the immune deficiency that arose from excess 2'-deoxyguanosine in the blood, researchers attempted to develop an inhibitor of the enzyme purine nucleoside phosphorylase (PNP). Their strategy was to develop a stable transition state analog for PNP. The compound they ultimately developed, DADMe-Immucillin-H, bound to PNP more than two million times more tightly that the substrate and was a powerful enzyme inhibitor (Schramm, V. L. 2007, *JBC*, 282: 28297).

Experiments like these have repeatedly demonstrated that stable transition state analogs can be powerful enzyme inhibitors. In doing so, they provide confirmation that our understanding of how enzymes catalyze reactions is correct. This understanding is a valuable aid to pharmaceutical development. Traditionally, the development of enzyme targets by pharmaceutical companies was a painstaking process of screening hundreds of thousands of compounds for those that had some activity against the enzyme. Once the first round of screening was done, teams of chemists made many variants of the compounds that showed some activity. These variants were again screened for activity, and it was hoped that some of the variants showed greater activity. This process was repeated over and over until an effective drug is obtained. However, by instead using the stable transition state theory as a guide, high-specificity and high-affinity inhibitors have been achieved with the synthesis of only a few molecules. This is unprecedented in drug design.

B. Test 2: Catalytic Antibodies

The effectiveness of stable transition state analogs provided strong support for the idea that enzymes catalyze reactions by binding most tightly to the transition state. We can take this idea a step further in the development of catalytic antibodies (abzymes). We can start with the generalization that enzymes are proteins that bind tightly to the transition state. If we could make our "own" proteins that bind tightly to a transition state, then would we not also expect that our protein would be an enzyme? It turns out that we can accomplish this by using a stable transition state analog as an antigen for immunization similar to making a vaccine. Typically, if an antigen were injected into an animal, it would trigger an immune reaction. In our case, the animal's immune system would then produce antibodies that bind tightly to the stable transition state analog. An antibody that binds to a stable transition state analog fits the generalization above that an enzyme is a protein that binds tightly to the transition state. As might have been predicted, these antibodies turn out to have enzymatic activity.

3 Extending the Model of Enzyme Catalysis to Include Coupling of a Reaction to an Energy Source

A. Variations of Our Model Are Informative

Now, let's consider a variation on our model. Before, we considered the case were the enzyme bound to the transition state where the transition state is approximately in the middle between

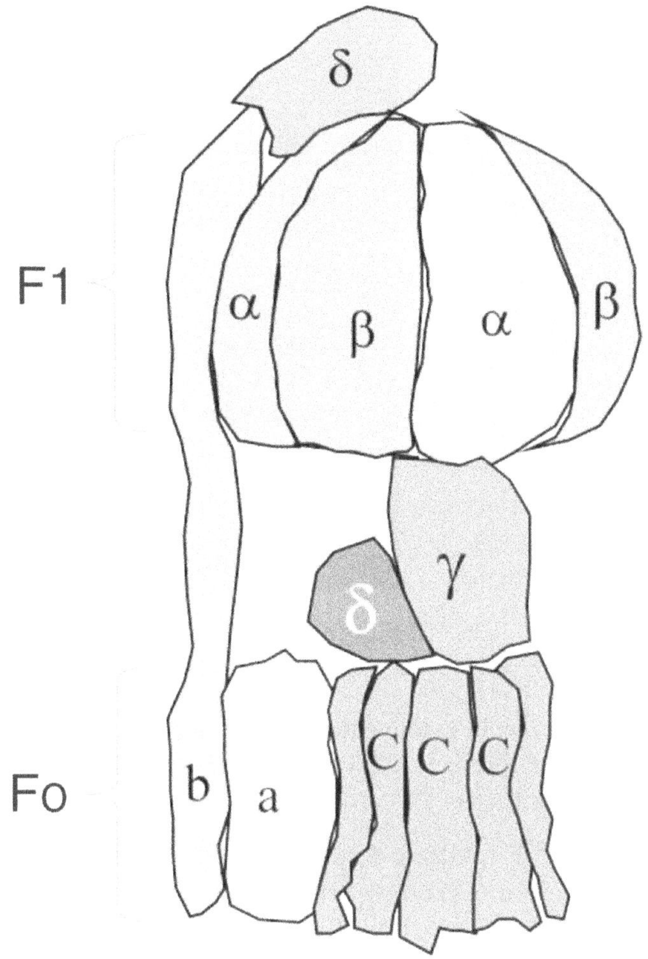

Figure 5.5. ATP synthase (also known as the F1/Fo ATPase). ATP synthase is a molecular machine composed of two main components. The first is Fo (o = oligomycin-sensitive component), which consists of the C and "a" subunits. The C subunits form a rotary motor in the mitochondrial inner membrane that is driven by the flow of protons (H^+) through the membrane via the C and "a" subunits. The rotation of the C subunits drives the rotation of the of the connected γ shaft that protrudes into the center of the α, β subunits. The F1 component consisting of the α, β subunits can also function as a motor, since it can drive rotation of the γ shaft by the hydrolysis of ATP. In the production of ATP, each of the β subunits of the synthase binds either ADP and a phosphate (Pi) or ATP. The synthesis of ATP from ADP and Pi is accomplished through binding of ADP and Pi, which progressively forces the phosphate to form a bond with ADP. It is binding that provides the energy to form ATP. Energy from the rotating γ shaft then distorts the β subunit, forcing the release of ATP.

Source: M. Poenie.

reactant and product. Under these circumstances the transition state is short-lived and rapidly decomposes into products that are released. What if the transition state were closer in form to the product? In this case the enzyme would bind the reactants and through binding, squeeze the reactants into products. The problem at that point would be how to get the products off the enzyme because the enzyme binds to them so tightly. This would require an input of energy where the energy it takes to release the product is at least equal to the amount of energy needed to make the product.

A beautiful example of this is seen in ATP synthase or the F1/Fo ATPase of mitochondria. This nanomachine is what makes most of our ATP in the mitochondria. The Fo component stands for oligomycin-sensitive component because the drug oligomycin binds to this part of the ATPase and blocks its function (see Figure 5.6). A major component of Fo segment of the ATPase are the ring of rotating C subunits which are driven to turn by a gradient of hydrogen ions. As it turns so also does the connected gamma shaft that goes up through the hollow center of the ring of alpha-beta subunits which constitute the F1 portion of the ATPase. A molecule of ATP is synthesized when ADP and phosphate are bound by the β subunits and effectively squeezed into ATP. At this point, the beta subunit is bound tightly to ATP and does not want to let go of it. Getting ATP off the beta subunit is achieved by the rotating gamma shaft that essentially forces the beta subunit to let go of the ATP.

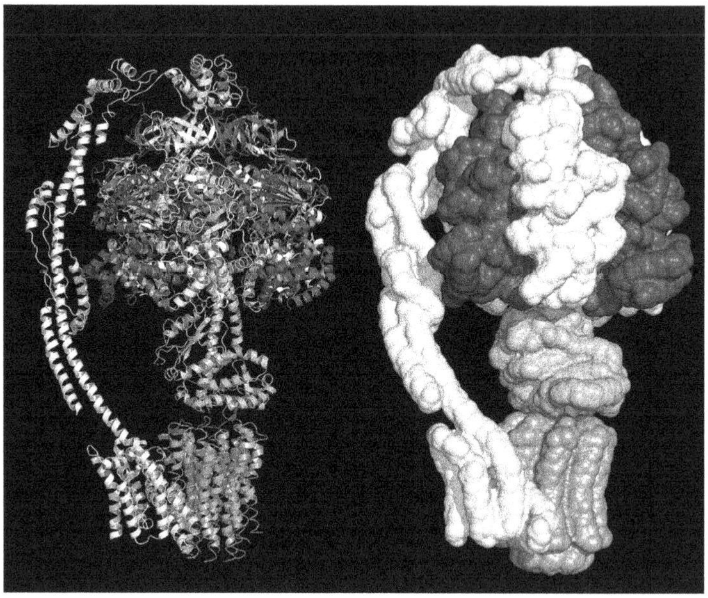

Figure 5.6.

Our understanding of how the ATP synthase actually makes ATP was originally worked out by Paul D. Boyer (1989, 2002), and John E. Walker (1998) who jointly received the Nobel prize in 1997 for elucidating the function and structure of ATP synthase. Boyer described the "binding change" mechanism in which the three β subunits of the F1 region exist in three different states that he called open, loose, and tight. However, these states are constantly shifting because they depend on the position of the rotating gamma shaft. The particular state each β subunit is in depends on the position of the rotating shaft. The γ shaft is bent, not symmetrical, and as it turns it bumps into one of the β subunits while at the same time it is farthest away from one of the other β subunits. The interaction with the γ shaft causes the β subunit to go into the open state, where ATP is free to leave. At the same time, the β subunit in the loose state binds ADP + Pi. Finally, the subunit farthest from contact with the shaft goes into the tight state. This is where ADP + Pi are converted into ATP. The rotating γ shaft next interacts with the β subunit in the tight state. It now converts to the open state, releasing ATP (see Figure 5.7).

1) Enzyme Cofactors

Enzymes are proteins made from amino acids. The twenty amino acids used in the assembly of proteins are limited in the kinds of chemistry they can perform. In order to extend the kinds of chemical reactions that enzymes can do, various specialized cofactors are used. A common example is in oxidation and reduction reactions. In these reactions, electrons are removed from one molecule and transferred to another. The molecule that loses the electrons is oxidized, whereas the one that receives the electrons is reduced. These reactions are used extensively in the production of energy in the form of ATP and in synthesis of complex molecules from simpler ones.

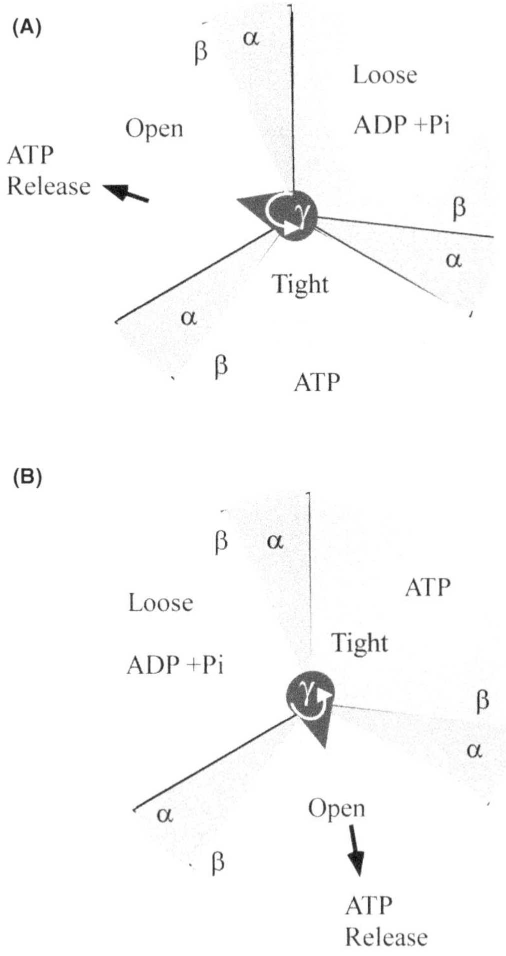

Figure 5.7. Boyer's Binding Change Model of ATP synthesis. In this figure the perspective is looking at the top of the ATP synthase F1 component. In the center is the rotating γ shaft (dark blue). Each β subunit is represented by a different color. In A, the gamma shaft is interacting with the green β subunit, forcing it into the open form. In the open form, the subunit releases ATP. Meanwhile, the blue β subunit is now in the tight state, where it is tightly bound to ATP. The pink β subunit is in the loose state, where it binds ADP and Pi. As the shaft rotates, the situation changes to that shown in B. Now the γ shaft is distorting the blue β subunit, causing it to go into the open state, where it releases ATP. The pink β subunit now goes into the tight state, where ADP + Pi are converted to ATP. Finally, the green β subunit now is in the loose state, where it binds to a new ADP + Pi.

Source: M. Poenie.

There are two prominent cofactors used in electron transfers: nicotinamide adenine dinucleotides (NAD+/ NADH; NADP+/NADPH) and the flavins (riboflavin). NAD+ is the oxidized form and NADH is the reduced form. NAD and NADP are virtually identical except for the additional phosphate in NADP, but NAD is normally used in energy metabolism whereas NADP is used in synthetic pathways and to keep the cytoplasm and to reduce glutathione which maintains the reducing environment of the cytoplasm.

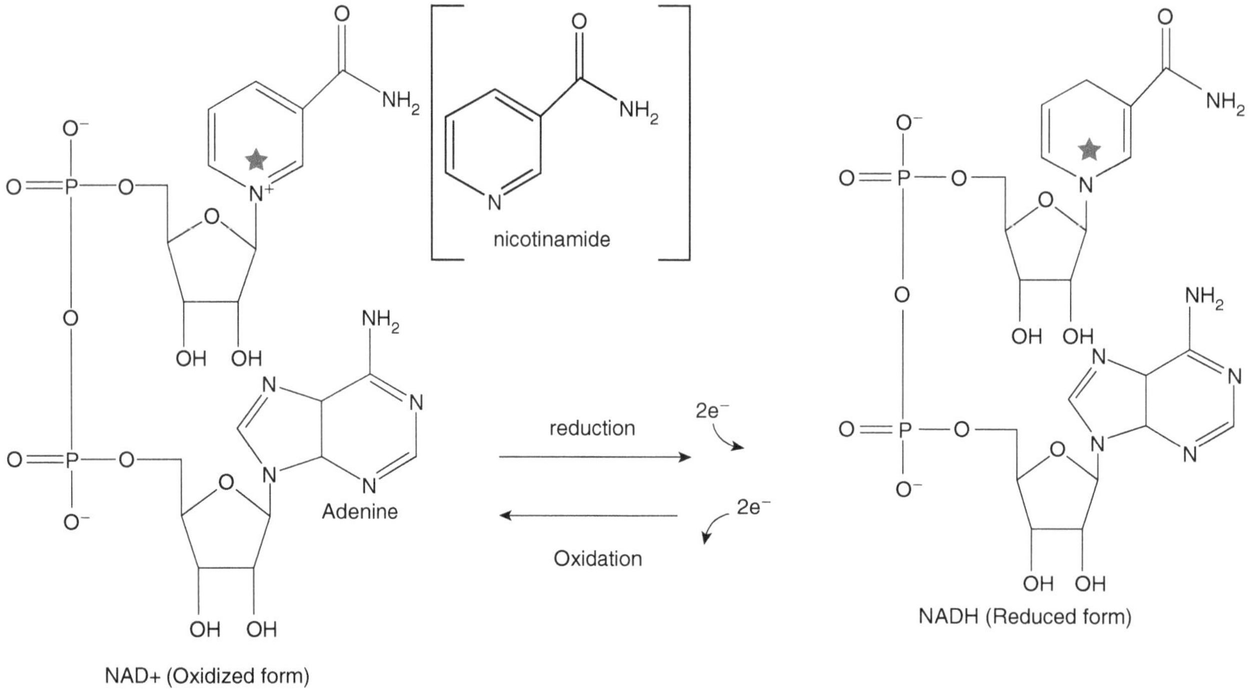

Figure 5.8. Nicotine adenine dinucleotide (NAD) is an electron carrier that undergoes reversible oxidation and reduction. One can think of the ribose and adenine portion of the molecule as a handle that is recognized by enzymes whereas the nicotinamide ring is the business end of the molecule that receives or donates electrons. The NAD+ form is the oxidized form, which can receive two electrons to become NADH. The focus of the two-electron reduction is the nicotinamide ring nitrogen (see the star). NADH can in turn transfer these electrons to other electron acceptors and go back to the oxidized state.

Source: M. Poenie.

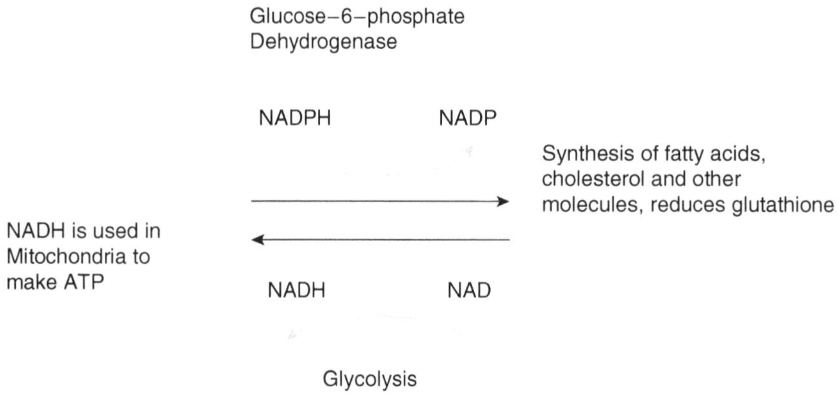

Figure 5.9. NAD+/ NADP+ and NADH/ NADPH are almost identical electron carriers, differing only by the phosphate on the ribose. The presence of a phosphate on the ribose distinguishes which enzymes will bind to the handle. Enzymes associated with synthetic pathways recognize the NADP form, whereas enzymes associated with breakdown pathways (typically for energy generation) recognize the unphosphorylated handle.

Source: M. Poenie.

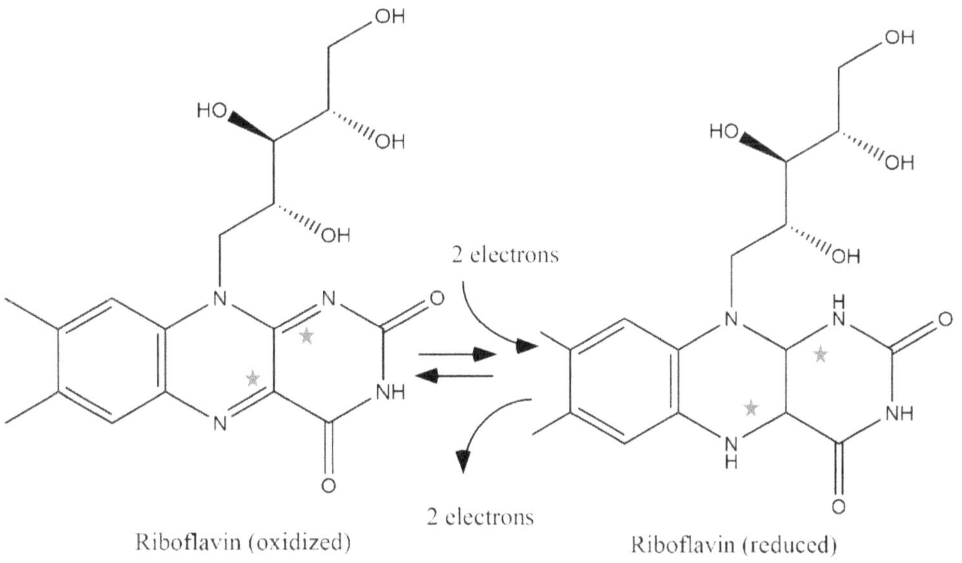

Figure 5.10. Flavin oxidation and reduction. Flavins such as riboflavin undergo reversible oxidation and reduction of their double bonds (see stars). In doing so they can carry electrons from one place in the cell to another.

Source: M. Poenie.

4 Enzyme Regulation

A. Regulatory Domains

Enzymes can be regulated in a wide variety of ways. We cannot discuss all of them here but we will offer a few examples. One rather common mode of regulation involves interaction between domains. It is common for enzymes to have a regulatory and a catalytic domain. One example is where the regulatory domain folds over the catalytic site and binds (see Figure 5.11). Often, it binds because the regulatory domain contains a pseudosubstrate site. A pseudosubstrate site is a stretch of amino acids that bind like the real substrate but cannot be acted on by the catalytic site. Instead, the pseudosubstrate binds to the catalytic site and just stays there preventing the catalytic site from acting on other molecules. In order to remove the pseudosubstrate, something else must bind to the regulatory domain so that it no longer binds to the catalytic domain. This, as shown in Figure 5.11.

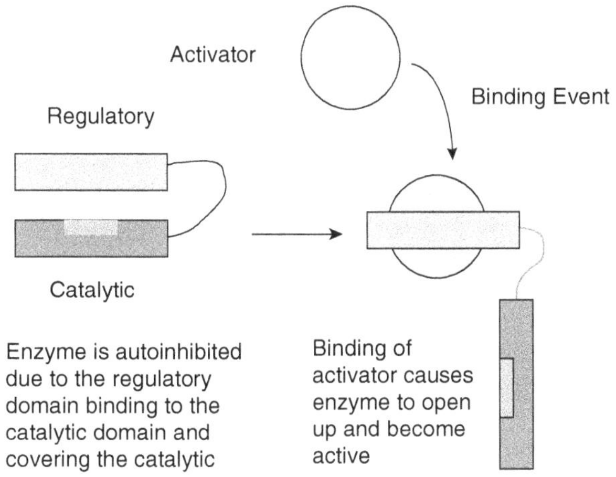

Figure 5.11. In this example of regulation, the enzyme has a regulatory and catalytic domain. The regulatory domain blocks the function of the catalytic domain.

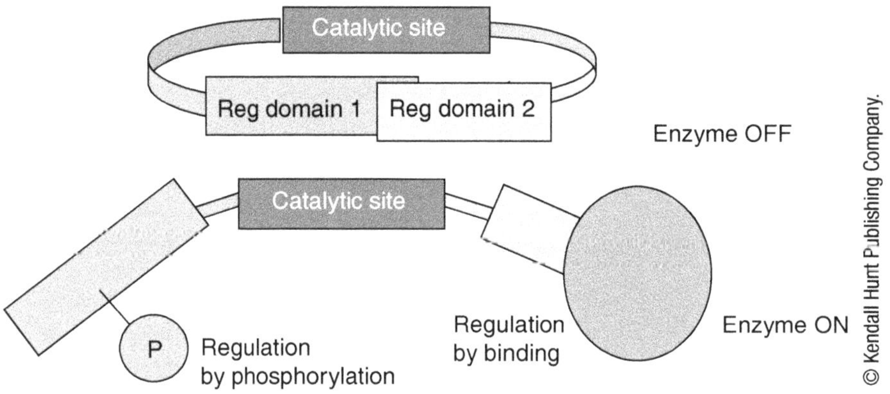

Figure 5.12. In a second example the enzyme has two regulatory domains that bind to each other and in doing so block the catalytic site. The enzyme can not could be activated by causing the regulatory domains to come apart. This could be caused by phosphorylation of one of the regulatory domains or binding of another molecule to one of the regulatory domans.

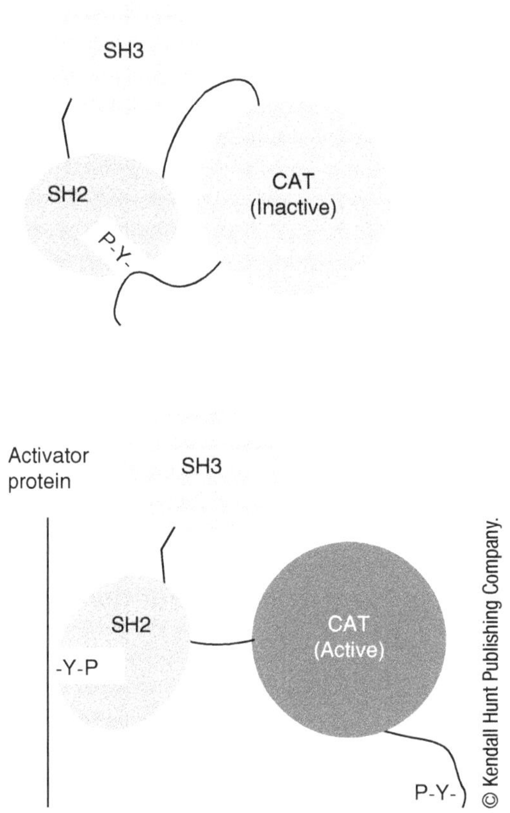

Figure 5.13. A third example is seen in the activation of Src. Src is a nonreceptor tyrosine kinase that is activated when it binds to tyrosine phosphates (Y-P) on an activator protein. The SH2 domain binds to tyrosine phosphate and adjacent amino acids. Normally Src is kept inactive because of a phosphorylated tyrosine near the N terminus of Src. When the Src SH2 domain binds to this tyrosine phosphate, it distorts the catalytic domain, making it inactive. When particular tyrosine phosphates on another protein are presented to Src, the Src SH2 domain binds preferentially to it causing Src to open up and become active.

B. Regulating Enzymes by Changing Their Location

In many cases, activation of an enzyme or other function involves changing its location. An example is seen with the insulin-activated glucose transporter known as GLUT4 or the Na / H+ ATPase of parietal cells that pump acid into the stomach. The glucose transporter is stored in vesicles underneath the plasma membrane. Since GLUT3 needs to be in the plasma membrane in order to let glucose into the cell, it cannot transport glucose while these vesicles are in the cytoplasm. Insulin binding to the insulin receptor causes these vesicles to fuse with the plasma membrane. Now, the transporter can let glucose into the cell. A similar situation is seen with the proton pumps of parietal cells in the stomach. When these cell are not secreting acid, the pumps are stored in vesicles underneath the plasma membrane. When these cells are stimulated to secrete by histamine or acetylcholine, the vesicles fuse with the plasma membrane. Now the pumps can function to secrete acid into the stomach. While only show a couple of examples, we shall see that regulation of function through a change in location is a common theme in cell biology.

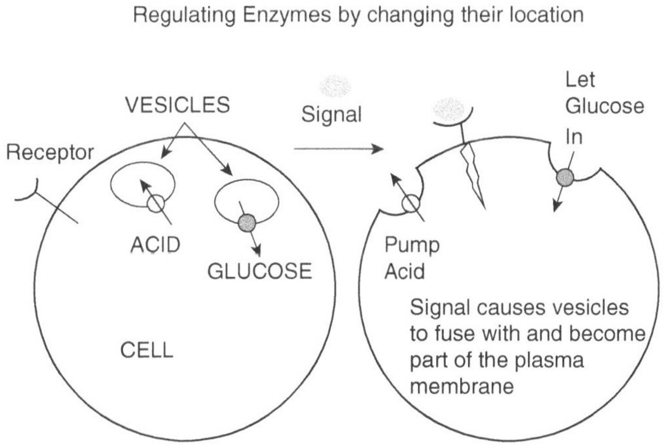

Figure 5.14. Enzyme regulation through a change of location. The illustration here shows regulation of two different enzymes stored in vesicles inside the cell. One is an acid pump that produces stomach acid, and the other represents the insulin-dependent glucose transporter. A signal on the outside of the cell (e.g., insulin or histamine) binds to a receptor, which generates a signal inside the cell, which in turn causes vesicles to fuse with the cell membrane. When this happens, the glucose transporter is able to let glucose into the cell while the acid pump moves H+ into the stomach.

References

Boyer, Paul D. 1989. "A Perspective of the Binding Change Mechanism for ATP Synthesis." *FASEB Journal* 3 (10): 2164–78.

Boyer, Paul D. 2002. "A Research Journey with ATP Synthase." *The Journal of Biological Chemistry* 277 (42): 39045–61.

Schramm, V.L., *Enzymatic transition state theory and transition state analogue design*. J Biol Chem, 2007. **282**(39): p. 28297–300.

Walker, John E. 1998. "ATP Synthesis by Rotary Catalysis (Nobel Lecture)." *Angewandte Chemie (International ed. in English)* 37 (17): 2308–19.

Lipids

6

Lipids are molecules that are typically composed of one or more hydrocarbon chains that make the molecule characteristically hydrophobic. In many cases, however, the lipid molecule can also have a hydrophilic component making the molecule amphoteric. The term *amphoteric* refers to having two natures, which in this case are the hydrophilic and hydrophobic components.

Neutral versus polar lipids

Some lipids are referred to as neutral lipids because they have no charged groups as part of the lipid. Polar lipids tend to have one or more charged groups.

1 Neutral Lipids

These include the following:

A. Triacylglycerols (triglycerides), which are used to store energy

B. Lipids built from isoprene subunits

1. **Steroid Lipids**
2. Prenyl lipids (geranylgeraniol, farnesol, dolichol, retinol)

Lipids built from isoprene subunits all go through an enzymatic step mediated by mevalonyl CoA reductase. This enzyme is blocked by atorvastatin and a wide variety of other drugs that reduce cholesterol.

A. Triacylglycerol and Fatty Acids

Triacylglycerols are the components of fats and oils and are made by the bonding of fatty acids to glycerol. Triacylglycerols are used primarily as a way to store energy in a concentrated form. They are present in many cells as small droplets and in adipose cells as a large droplet.

Both triacylglycerol and a majority of membrane lipids incorporate fatty acids into their structures. Fatty acids have for the first carbon a carboxyl or carboxylic acid group followed by a long hydrocarbon chain. Fatty acids differ in the length of the hydrocarbon chain and the number and type of double bonds. By far the most common type of double bond is known as a cis double bond. The opposite of cis is a trans double bond. Perhaps you have heard of trans fats. This refers to fatty acids that contain trans double bonds, usually arising from the hydrogenation of vegetable oils. Whether fatty acids have no double bonds or cis or trans double bonds has a big impact on the property of the lipid.

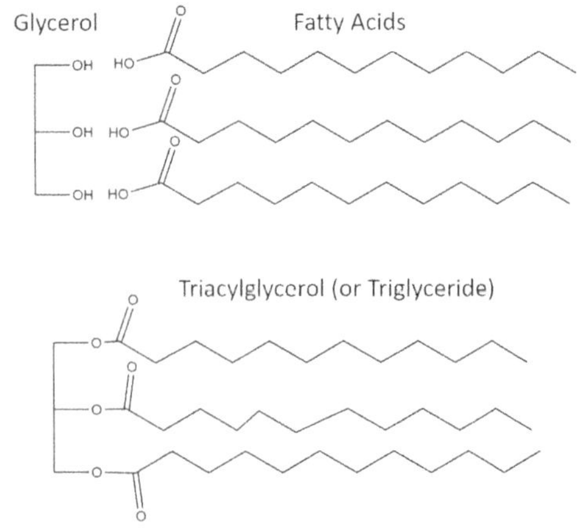

Figure 6.1. Triacylglycerols are lipids that are used primarily as a means for storing energy in a concentrated form. These compounds are formed by bonding three fatty acids to glycerol.

Source: M. Poenie.

Figure 6.2. Liver cells stained with Oil Red O to reveal lipid droplets.

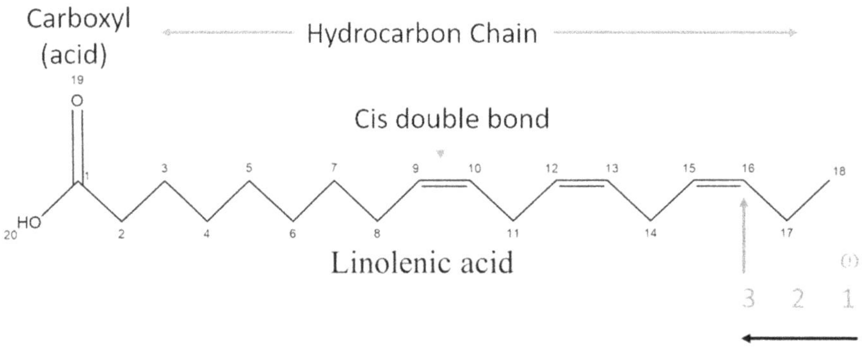

Figure 6.3. Linolenic acid is an example of a fatty acid. Fatty acids have an acidic carboxyl group at one end followed by a hydrocarbon tail. In the case of linolenic acid, the total number of carbons is eighteen. Some fatty acids have double bonds in the chain, and these are normally cis double bonds (see Figure 6.4). In the case of linolenic acid there are three cis double bonds. Cis double bonds determine how fatty acids pack together. While chemists (IUPAC naming) name the carbons, beginning with the carboxyl carbon, it has become popular to count from the last carbon (the omega carbon) back to the beginning of the first double bond. In linolenic acid the first double bond occurs at carbon three, so this is an omega-3 fatty acid. There are some fatty acids that we do not make and that must be supplied in the diet. Omega-3 fatty acids cannot be made by mammals, although linolenic acid can be used to produce the period longer chain omega-3 fatty acids.

Source: M. Poenie.

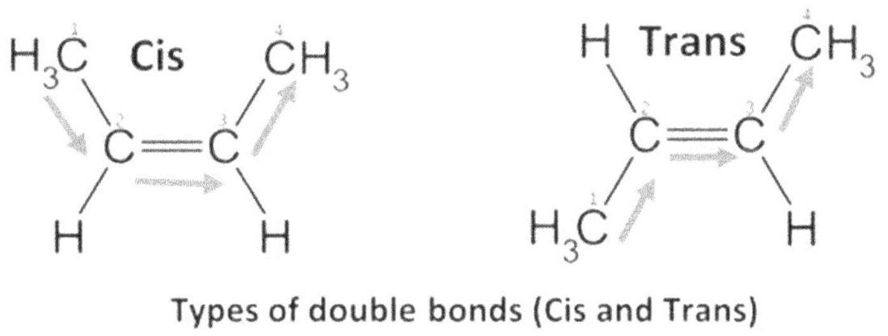

Figure 6.4. Cis and trans double bonds. Double bonds in fatty acids can be either cis or trans, although trans double bonds are normally rare. The terms "cis" and "trans" refer to the relationship between the double bond and the carbon chain. If the carbon chain "enters" and "leaves" on the same side of the double bond, it is a cis double bond. If the carbon chain enters on one side of the double bond and leaves on the other side, it is a trans double bond.

Source: M. Poenie.

The kinds of fatty acids in a triacylglycerol determine if the lipid is solid or liquid at room temperature. The presence of double bonds in fatty acids tends to give triacylglycerols that are liquid at room temperature. This is because cis double bonds create a bend in the hydrocarbon chain that prevents them from packing together and bonding. In the absence of double bonds, the hydrocarbon chains can pack together and form Van der Waals bonds, resulting in a lipid that is solid at room temperature. This difference can be readily seen by comparing vegetable oil to vegetable shortening.

Figure 6.5. Vegetable oil is liquid at room temperature because many of the fatty acids have cis double bonds. In the case of Crisco and other brands of vegetable shortening, the double bonds have been removed by hydrogenation. The resulting triacylglycerols are solid rather than liquid at room temperature.

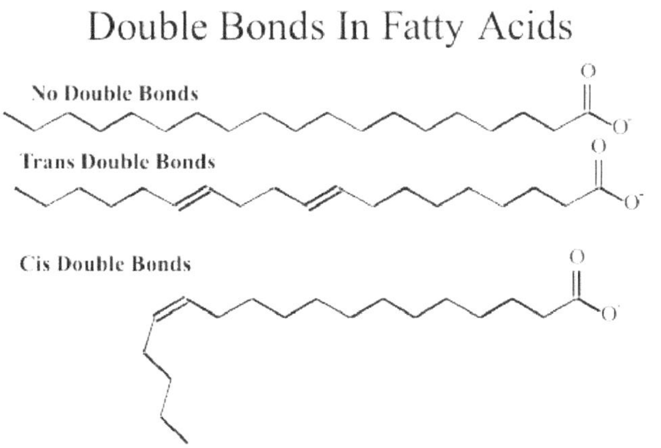

Figure 6.6. A comparison of fatty acids having no double bonds, trans double bonds or cis double bonds.

Source: M. Poenie.

		Double Bonds				
		0	1	2	3	4
Number of carbons	C12	Lauric				
	C14	Myristic				
	C16	Palmitic	Palmitoleic			
	C18	Stearic	Oleic	Linoleic	Linolenic	
	C20	Arachidic				Arachidonic

Figure 6.7. Fatty acid characteristics. Fatty acids vary by the length of the carbon chain and the number of double bonds. This is often expressed as chain length: number of double bonds. For example, linoleic acid is 18:2. Both chain length and double bonds make important contributions to the properties of lipids and membranes.

Source: M. Poenie.

B. Steroids

Steroids are another type of neutral lipid that are synthesized from small subunits called isoprene. All steroid lipids including cholesterol and steroid hormones contain what is called the steroid nucleus (shown in green in Figure 6.8). Cholesterol is an important structural part of the plasma membrane. Steroid hormones being hydrophobic are able to pass through the cell membrane and bind to receptors in the cytoplasm or nucleus in order to trigger the transcription of particular genes.

Cholesterol, steroids, and many other lipid based molecules are all derived from the mevalonate/isoprene pathway. One of the early steps involves the reduction of HMG-CoA to give mevalonate. This pathway is blocked by statins such as atorvastatin, drugs that are used to reduce cholesterol levels. Of course, they do not just reduce cholesterol.

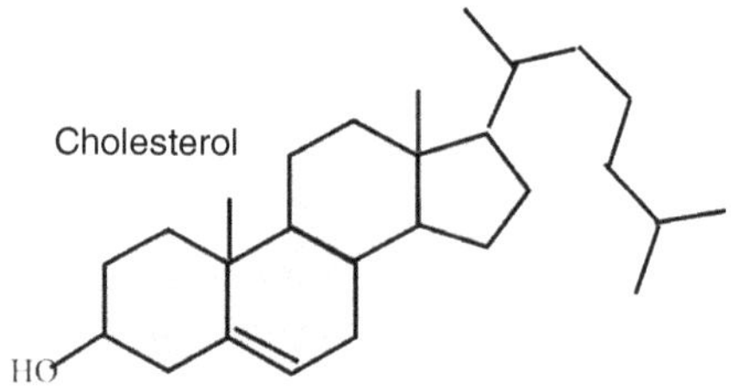

Figure 6.8. Cholesterol is an example of a steroid. All steroids have the steroid nucleus, shown here in green.

Source: M. Poenie.

C. Protein Lipidation

Lipids can be attached to proteins to make them associate with lipid membranes. There are two types of protein lipidation that are common in cells. They are fatty acid acylation and prenylation.

In fatty acid acylation, fatty acids are attached to proteins on an N-terminal glycine or on cysteine. Myristic acid (14:0) can be attached to the N-terminal glycine and palmitic acid (16:0) can be attached to the sulfur of a cysteine amino acid. In the case of adding palmitic acid to a cysteine sulfur, the fatty acid is readily removed. In prenylation, either farnesol or geranylgeraniol (products of the mevalonate/isoprene pathway) are attached to the sulfur of a cysteine, but in this case they are attached through an ether linkage and are not removed. In order for this attachment to occur, the protein is first cleaved at a cysteine amino acid, leaving a C-terminal cysteine. The lipid is then attached and finally the cysteine carboxyl is methylated.

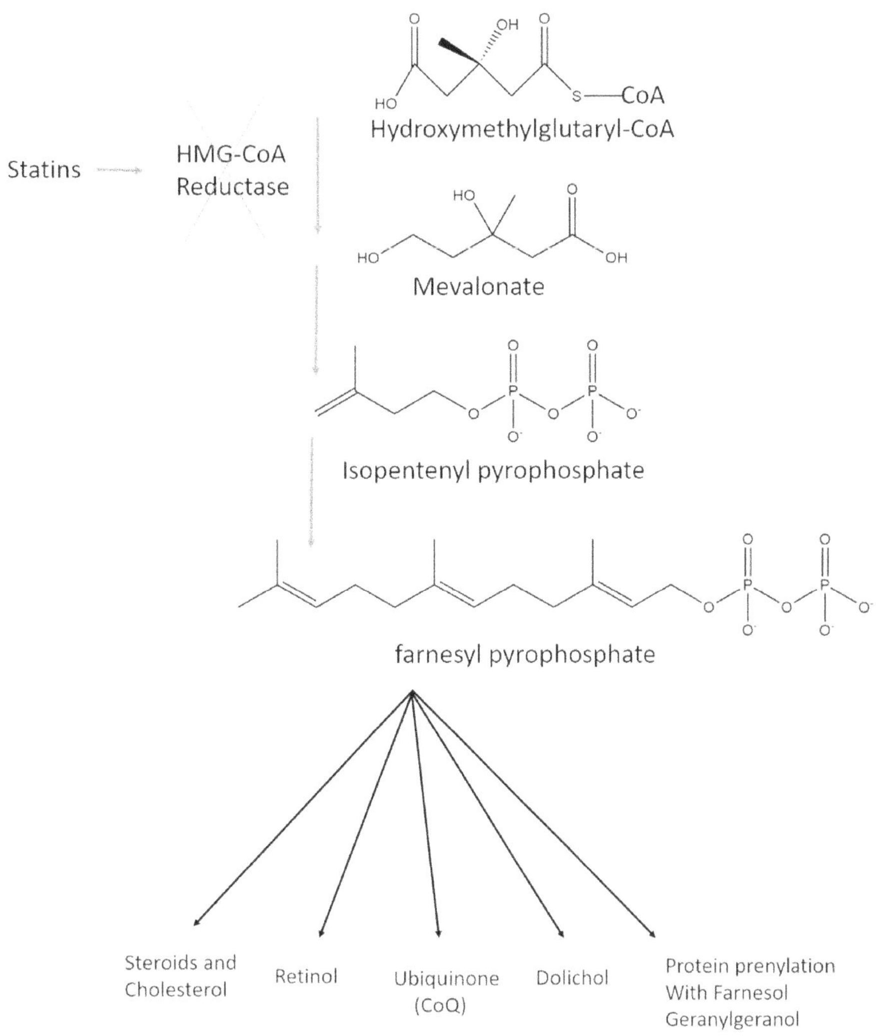

Figure 6.9. The pathway for making steroids is used to make a number of other important lipids. It begins with the reduction of hydroxymethylglutaryl-CoA to mevalonate. From here chains are built up by adding successive five carbon units derived from isopentenyl pyrophosphate. Statin drugs which block synthesis of cholesterol also block the production of a variety of other lipids important in cell function.

Source: M. Poenie.

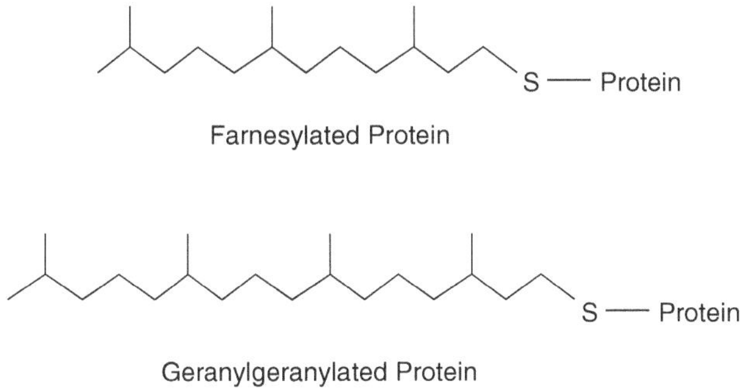

Figure 6.10. The addition of farnesol or geranylgernol to a protein causes the protein to associate with the plasma membrane. Without this lipid the protein does not go to the right location and does not function properly.

A key difference should be noted between acylation and prenylation that we will discuss later. Prenylated proteins are targeted to the phospholipid bilayer, whereas acylated proteins are targeted to lipid rafts.

2 Building Membranes with Polar Lipids: Glycerophospholipids (Phosphatides) and Sphingolipids

A. Glycerophospholipids

Cell membranes are mainly composed of polar lipids where one portion of the lipid molecule is hydrophobic and another portion is hydrophilic. Molecules that have this characteristic are call amphipathic because they have two natures (see Figure 6.11). The cell membrane is largely composed of three types of lipids: The glycerophospholipids, the sphingolipids and cholesterol, all being at least somewhat amphipathic.

The glycerophospholipids (phospholipids) are one of major groups of lipids in the membrane. The most basic of the phospholipids is phosphatidic acid. It has two fatty acid chains esterified to glycerol, and the third carbon is esterified to phosphoric acid (Figure 6.11). In fact, there is normally very little phosphatidic acid in the membrane although it can be made as a short lived signal due to the action of certain phospholipases (see below). Most phospholipids are further modified to have what is called a head group linked to the phosphate of phosphatidic acid. The major head groups are choline, ethanolamine, inositol and serine (see Figure 6.12). Normally, phospholipids have a "head group" attached to the phosphate. The head group distinguishes the different phospholipids.

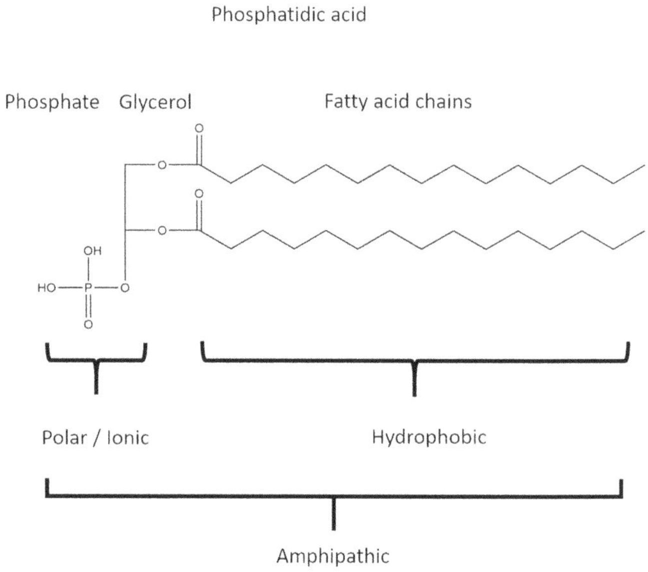

Figure 6.11. Phosphatidic acid. All the glycerophospholipids used in making membranes have phosphatidic acid at their core. Because of that, the names of all the membrane phospholipids begin with "phosphatidyl-." Various head groups attached to the phosphate then differentiate the membrane phospholipids. These phospholipids have a polar/ionic side and a hydrophobic side. Molecules with both a polar/ionic character and a hydrophobic character are referred to as amphipathic (Greek = two natures).

Source: M. Poenie.

B. Polar Lipids Spontaneously Organize When Put in Water

The amphipathic character of polar lipids, including the phospholipids, is to organize so that the polar end of the molecule is in water and the hydrophobic end is out of or away from the water. Structures that allow

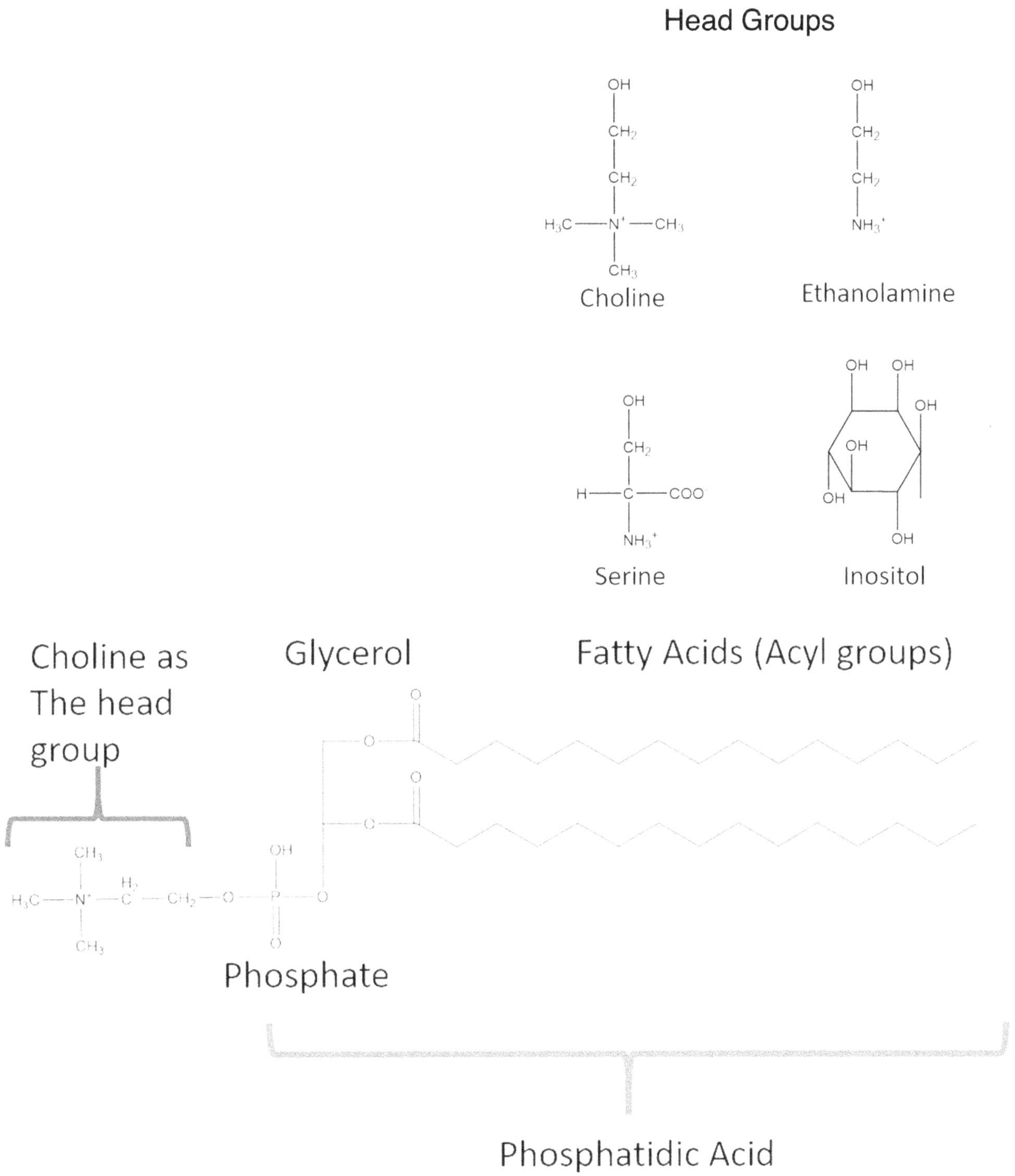

Figure 6.12. Membrane phospholipids. There are four different membrane phospholipids, which differ based on the head group that is attached to phosphatidic acid. Each is a derivative of phosphatidic acid. One example shown below has choline as the head group, making this lipid phosphatidyl choline.

Source: M. Poenie.

for this behavior are micelles, lipid vesicles (liposomes), and bilayers. Note that as drawn in Figure 6.13, a bilayer sheet still has the sides exposed to water. This is not really going to happen. Instead, the exposed hydrophobic region will always be minimized. This can be accomplished by having the edges curl up as seen in Figure 6.14.

As a general principle, polar lipids will try to find an arrangement that minimizes exposure of the hydrophobic portion of the molecule to water.

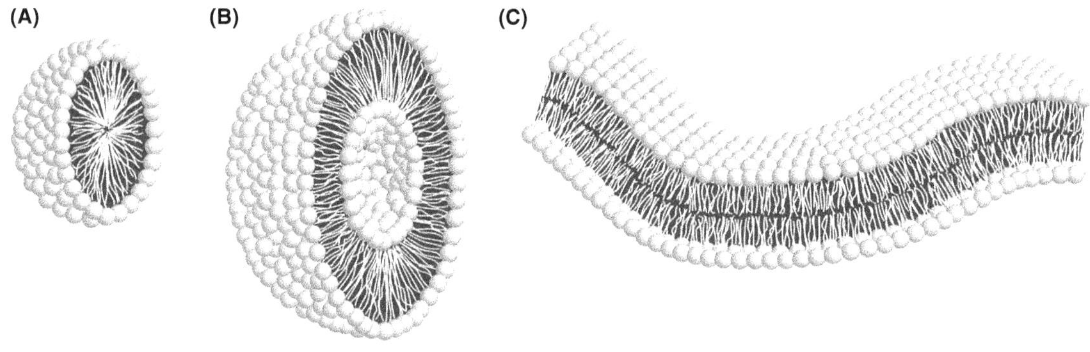

Figure 6.13. Lipids spontaneously form different types of structures in water. These are micelles **(A)**, liposomes **(B)**, and bilayers **(C)**. Each structure represents an attempt to maximize the exposure of hydrophilic parts of the lipid to water while minimizing exposure of hydrophobic components of the lipid molecules to water. Micelles have one layer of lipid molecules, with the hydrophilic head groups exposed to the water and the hydrophobic fatty acid chains in the interior. Liposomes are made from a double later of lipid molecules. The representation of bilayers is somewhat misleading because they appear here where the hydrophobic regions are exposed to water on the sides.

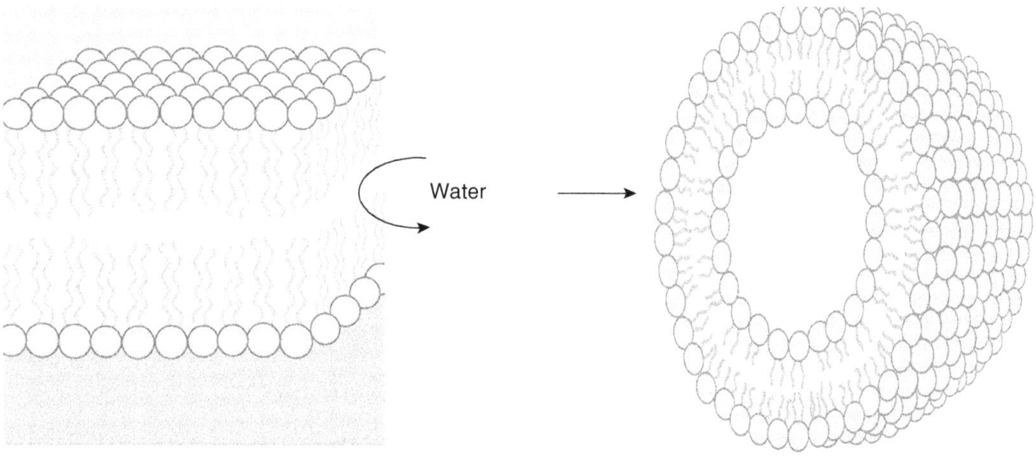

Figure 6.14. While a phospholipid bilayer might be shown as a planar structure like that on the left, that structure would not actually exist. The lipid chains are hydrophobic and will attempt to minimize their exposure to water. As a result, a planar structure like that on the left would rearrange so that the hydrophobic lipid chains are not exposed to water, giving rise to structures like the liposome shown on the right. The bilayer depicts the structure of the membrane enclosing larger structures like cells where regions of the membrane could be relatively flat.

C. Membrane Fluidity

The difference between vegetable oil and vegetable shortening shown previously is relevant to membranes which can similarly exist in a solid (gel) or liquid state depending on the temperature. Temperature controls the fluidity of the membrane through the motions of membrane lipids. The state of the lipids is often referred to as their phase whether solid or liquid. In the solid or gel phase, membrane lipids are packed together more closely and the fatty acid chains are aligned and more ordered. As the temperature increases, the fatty acid chains exhibit more motion and become disordered. These motions affect the fluidity of the membrane (Figure 6.15).

The fluidity of the membrane at any given temperature depends on the types of fatty acids in the lipids. Fatty acids influence the fluidity of the membrane through their degree of unsaturation and the length (number of carbons) of the fatty acid chain. The term saturation is based on the idea that double bonds are formed by removing hydrogens bonded to the carbons of the fatty acid. Conversely, by adding hydrogen to the fatty acids with double bonds, the double bonds are removed such that the carbon-carbon double bonds are replaced with carbon-hydrogen bonds. Thus, saturated fatty acids have no double bonds whereas unsaturated fatty acids have double bonds.

Saturated fatty acids pack together and bind to each other, giving the lipids a tendency to form a solid or gel state. Unsaturated fatty acids are prevented from packing together because cis double bonds create bends in the fatty acid chains.

Fatty acid chain length is another factor controlling the state of the lipids. Lipids with longer chain fatty acids have a tendency to go into the gel phase whereas shorter chain fatty acids cause the membrane to be more fluid. Changing the temperature can cause lipids to go from the gel phase to the liquid phase, which is referred to as a phase transition. We characterize the tendency of membrane lipids to be in a the gel or fluid state in terms of the melting temperature. Carrying over from our previous discussion, lipids with longer fatty acid chains and fewer double bonds will exhibit a higher melting temperature. Conversely, lipids with shorter the fatty acids chains and more unsaturated fatty acids will melt at a lower temperature.

The temperature where the membrane changes from solid (gel) to liquid is called the transition temperature (Figure 6.16).

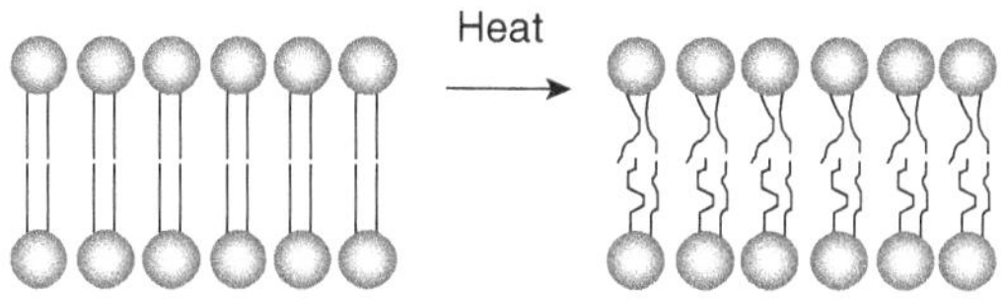

Figure 6.15. The fatty acid chains of membrane lipid molecules have a tendency to bond together due to formation of Van der Waals bonds. It is easier for the saturated fatty acid chains to align and form Van der Waals bonds compared to unsaturated fatty acid chains. Longer fatty acid chains also allow more Van der Waals bonds to form. When there are more bonds between the fatty acid chains, the lipids are more like a solid or gel (as in vegetable shortening). These Van der Waals bonds are weak and can be broken by increasing the temperature. When this happens, the fatty acid chains move more vigorously and the membrane becomes more fluid.

D. Membrane Fatty Acids in Plants and "Cold-Blooded" Animals

The transition temperature is a critical feature of cell membranes. If the membrane were actually to go into the gel state, it would kill the cell. In order to function properly, cells must maintain their membrane so that the lipids have a certain degree of fluidity. The membrane lipids must not freeze but they membrane cannot be too fluid either. This represents a serious problem for plants and poikilothermic (cold blooded) animals whose temperature is determined by the environment. In order to survive at different temperatures they must modify the composition of their membrane lipids. This process is called homeoviscous adaptation.

An example of homeoviscous adaptation is seen in a study of lizards who were kept at different temperatures. The lizards were raised in cold (LT), normal (NT), or hot (HT) environments. As can be seen in Table 6.1, the membrane lipids from lizards raised at different temperatures show a marked decrease in unsaturated fatty acid as the temperature of their environment increases.

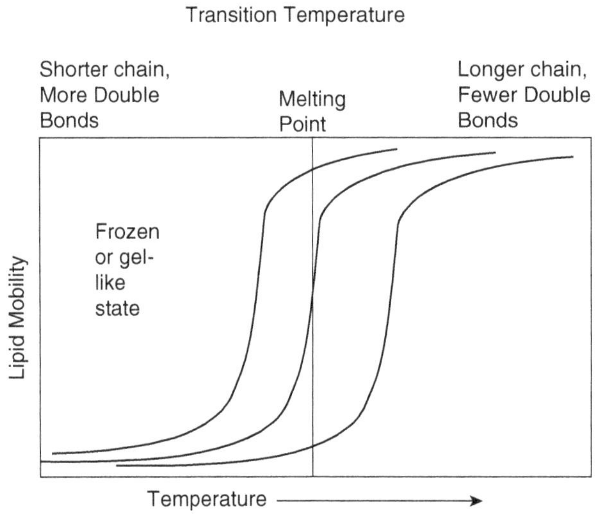

Figure 6.16. Membrane lipids exhibit a transition temperature where they change from a more solid state to a more liquid state. The transition temperature depends on the nature of the fatty acid chains in the lipid. Longer-chain fatty acids and more saturated fatty acids (fewer cis double bonds) give lipids that melt at a higher temperature. Lipids with shorter chains or more cis double bonds melt at a lower temperature.

Table 6.1. Changes in unsaturated fatty acids with temperature

	LT	NT	HT
18:2	25.7 ± 2.5	20.1 ± 1.7	19.4 ± 1.9
18:3	7.3 ± 0.4	1.4 ± 0.1	0.8 ± 0.1
20:3	0.1 ± 0.0	n.d.	0.1 ± 0.0
20:4	2.6 ± 0.2	2.4 ± 0.2	2.2 ± 0.1
20:5	0.7 ± 0.2	0.6 ± 0.1	0.4 ± 0.1
22:5	0.4 ± 0.0	0.3 ± 0.0	0.1 ± 0.0

Lizards were raised at cold (4°C), normal (25°C) or warm temperatures (37°C) and the fatty acid composition of their membrane lipids was analyzed. The data here show that as temperature increases, the amount of unsaturated fatty acids decreases.

Data from Shen, Li, and Gao (2005).

3 Sphingolipids

Another major group of membrane lipids are the sphingolipids. Sphingolipids are a family of about 4000 lipid species that are built using sphingosine rather than glycerol (Figure 6.17). Recent studies are revealing that these lipids are important in cell signaling and disease as well as being structural components of the membrane. Like the glycerophospholipids they are amphipathic with a polar end and a nonpolar end.

Sphingosine itself is not a structural part of membranes but it represents an important signaling molecule when it is phosphorylated. Addition of a fatty acid to the nitrogen on sphingosine gives a ceramide. Sphingosine itself has a long hydrophobic tail, and the fatty acids that are bonded to sphingosine tend to be long, ranging in length from 16 to 24 carbons and in special cases as long as 32 carbons. Ceramides are the starting point for building the many different sphingolipids.

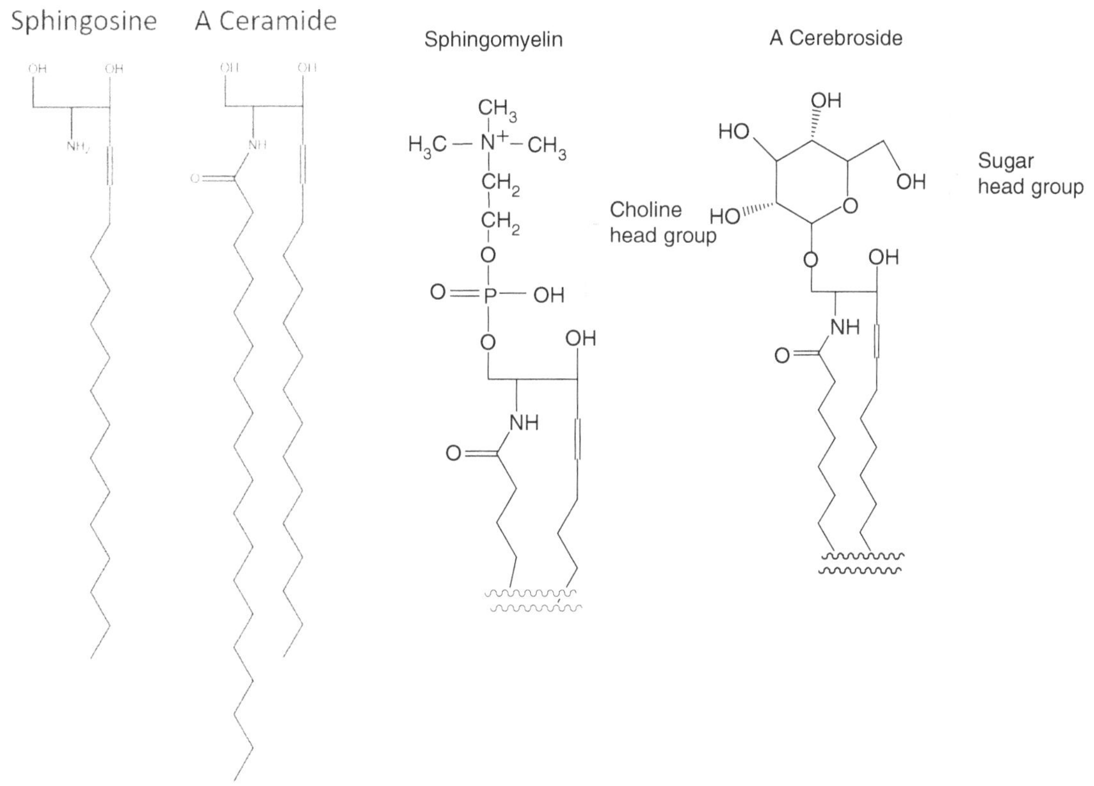

Figure 6.17. Sphingolipids. Sphingolipids are based on sphingosine and exhibit the same amphipathic character as seen in the glycerophospholipids. The amide linkage of a fatty acid to sphingosine gives a ceramide which is the hub for making the various family members of the sphingolipids. There are a variety of different additional molecules that can be attached to ceramide, similar to the head groups of the glycerophospholipids. These include choline, to give sphingomyelin, and simple sugars, to give cerebrosides. The addition of more complex sugars that contain N-acetylneuraminic acid (sialic acid) gives the gangliosides.

Source: M. Poenie.

Ceramides can be further modified by adding different kinds of molecules to the OH group at the one position on sphingosine, much the same as head groups are added to phosphatides. Phosphocholine can be added to the OH to give sphingomyelin. Alternatively, sugars can be added to give glycolipids. Glycolipids can be further divided into neutral and acidic sphingolipids depending on whether or not they contain sialic acid. If they do contain sialic acid they are called gangliosides. If not they are either cerebrosides if a single sugar is attached (usually galactose or glucose) or oligoglyocosylceramides if multiple sugars are attached.

A. Sphingolipids and ABO Blood Groups

The ABO blood group system is based on a particular configuration of sugars attached to sphingolipids (and to proteins of some people; Figure 6.18). Barring certain mutations, all people make the type O oligoglycosylceramide. Some people have an enzyme that adds N-acetylgalactosamine to the O oligoglycosylceramide and they are designated type A. Other people have an enzyme that transfers galactose instead of N-acetylgalactosamine. These are type B. Finally there are some people that have both enzymes and can transfer both types of sugars. These are type AB.

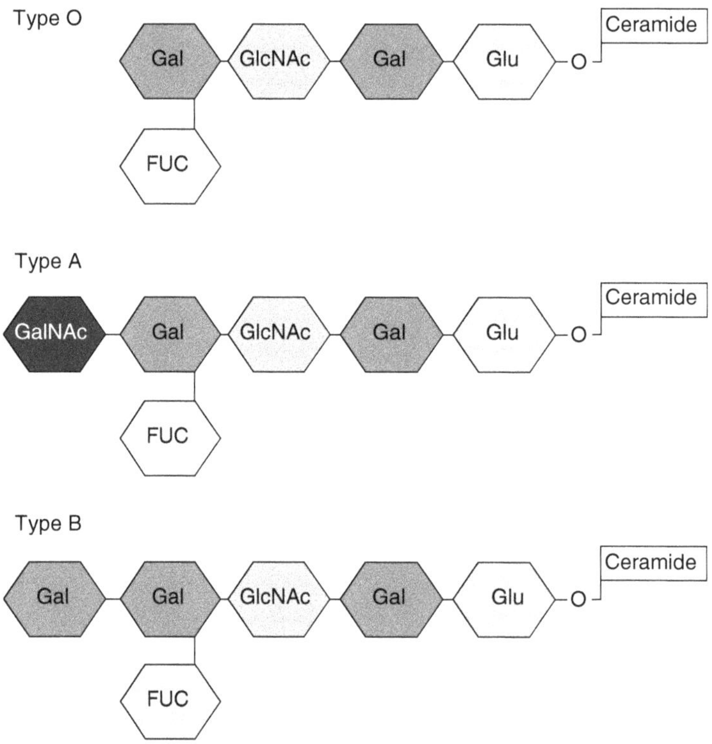

Figure 6.18. ABO blood groups. The ABO blood groups are based on a particular configuration of sugar chains added to a ceramide sphingolipid of red blood cells. The type O configuration is normally made by everyone. However, it can be further modified by the addition of either N-acetylglactosamine or galactose. Abbreviations Fuc Fucose, Gal Galactose, Glu Glucose, GlcNAc N-acetylglucosamine, GalNAc N-acetylgalactosamine

B. Sphingolipids and Lipid Rafts

Sphingolipids and cholesterol have been implicated in the formation of small regions within the plasma membrane of cells called lipid rafts. The idea is that sphingolipids preferentially associate with cholesterol to form regions that are thicker and more viscous than adjacent non-raft regions of the membrane. These lipid raft regions are thought to associate with different types of membrane proteins than the non-raft regions. We will discuss rafts more in the next chapter but it must be said that there is evidence for and against the idea of lipid rafts and in particular, whether cholesterol is a component of rafts. Lipid rafts have been implicated in cell signaling events, and in the sorting of proteins.

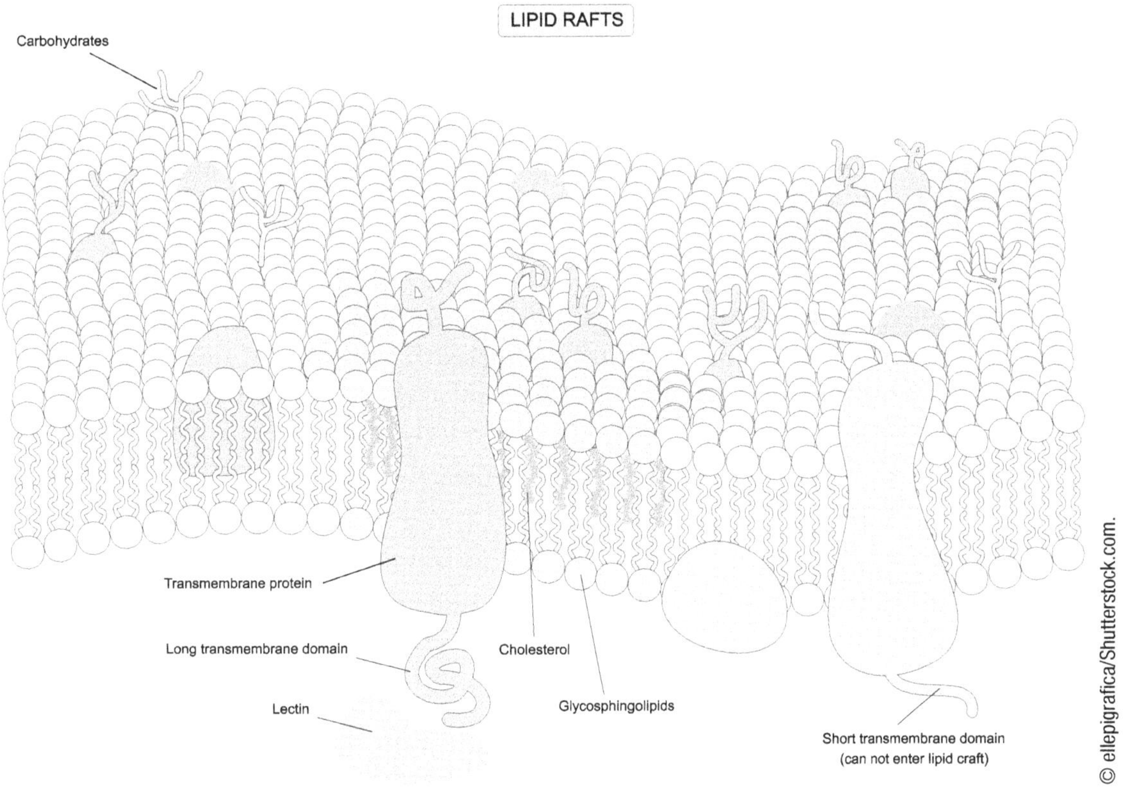

Figure 6.19. Lipid rafts. The membrane is made up primarily of glycerophospholipids (light brown), sphingolipids (purple) and cholesterol (red). The lipid raft model proposes that sphingolipids preferentially associate with cholesterol to form local regions within the membrane that have unique properties. They are thicker, more hydrophobic, and more viscous than the rest of the membrane.

4 Lipids as Signals

While lipids are structural components of membranes, they also play critical roles in cell signaling. Two of the key signaling components of the lipids are the fatty acids and the inositol head group. The inositol head group contains five OH groups that can be phosphorylated.

A. Phosphorylation

While lipids are essential components of the membrane, they also play critical roles in cell signaling. One way that lipids are used in signaling is through phosphorylation. The inositol ring of phosphatidylinositol can be phosphorylated at different positions on the ring. The position on the ring where a phosphate is added directs the cell into entirely different pathways. For example, Figure 6.20 shows phosphatidylinositol where the inositol has no phosphates. Next to it are examples where the inositol is phosphorylated either at the 4 and 5 positions of the inositol ring or at the 3,4, and 5 positions of the ring. These mean completely different things to the cell. The ring phosphorylated at only the 4 and 5 positions is a substrate for the enzyme phospholipase C which cleaves off the phosphoinositol ring. One the other hand, when the ring is phosphorylated at the 3,4, and 5 positions it becomes a ligand for a particular protein domain called a pH domain. Phosphorylations at different positions on the inositol serve as different signals. For example, phosphorylation at the 4 and 5 position on the inositol ring gives rise to inositol 1,4,5-trisphosphate when the head group is cleaved off the lipid. This is a signal that causes calcium release from the ER. On the other hand, phosphorylation at the 3,4, and 5 position means something completely different. In this case, it becomes a binding site for proteins containing PH domains.

B. Phospholipases

A second way that glycerophospholipids are used in signaling is through the action of phospholipases. There are different kinds of phospholipases that attack and cleave phospholipids at different positions. Phospholipases A1 and A2 cleave off acyl groups from the 1 position or 2 position respectively. Phospholipase C cleaves between the third glycerol carbon and the phosphate. Phospholipase D cleaves between the phosphate and the head group (Figure 6.21).

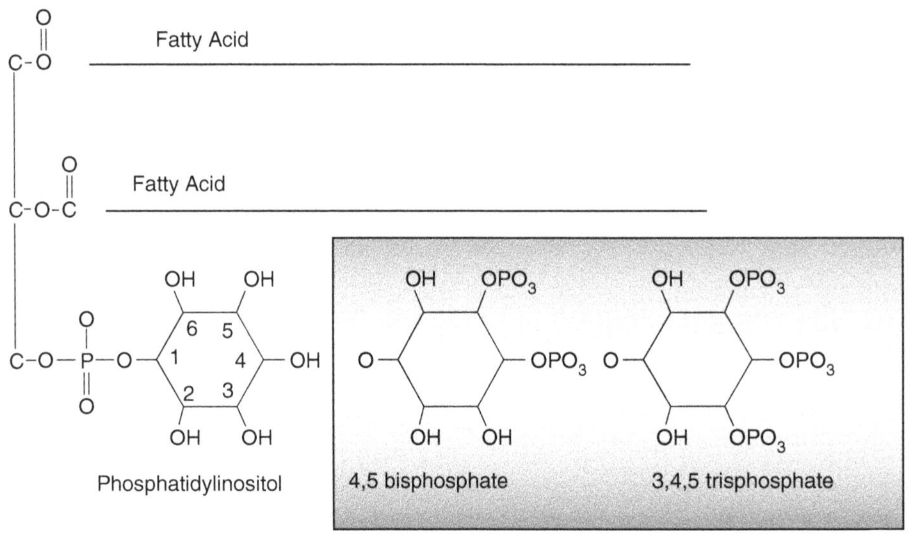

Figure 6.20. Phosphorylation of phosphatidylinositol. Phosphatidylinositol has the inositol ring for its head group. This ring can be phosphorylated at different positions which different phosphorylation states serve different functions. In the box, two different phosphorylation states are shown. Phosphorylation at the 4 and 5 positions gives phosphatidylinositol 4,5-bisphosphate. Phosphorylation at the 3,4, and 5 positions gives phosphatidylinositol 3,4,5-trisphosphate.

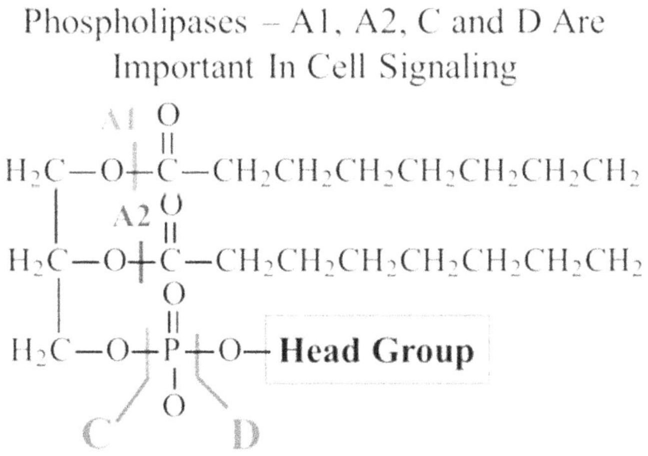

Figure 6.21. Different phospholipases cleave glycerophospholipids at different positions resulting in different cleavage products. Phospholipases A1 and A2 cleave off a fatty acid from either the first or second position of the glycerol. The result is a free fatty acid and a lysolipid. Phospholipase C cleaves between the third carbon and the phosphate resulting in a diacylglycerol and the phospho-head group. Phospholipase D cleaves between the phosphate and the head group resulting in phosphatidic acid and the free head group.

Source: M. Poenie.

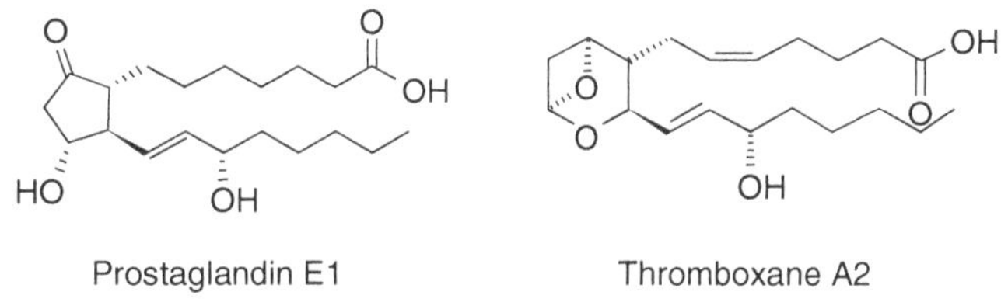

Figure 6.22.

Phospholipase A2 is particularly significant because the fatty acid called *arachidonic acid* (20 carbons, 4 double bonds) is most often located at the 2 position. Arachidonic acid is a substrate for an enzyme called *cyclooxygenase (COX)*. Cylcooxygenase is the first enzymatic step in converting arachidonic acid to *prostaglandins* and *thromboxanes*, lipid hormones or signals (Figure 6.22). Prostaglandins are a family of compounds that do many things, but often they regulate smooth muscles. Thromboxanes are related, but they most often function in inflammation.

Cyclooxygenase is inhibited by aspirin and other NSAIDs (nonsteroidal anti-inflammatory drugs). NSAIDs by inhibiting cyclooxygenase block the production of both prostaglandins and thromboxanes. There is yet a third group of lipid hormones, called HETE(s) for short, derived from arachidonic acid that we will not go into here.

C. NSAIDs and Fever

You are probably familiar with the fact that aspirin and other NSAIDs block fever. That is because prostaglandin E is part of the signaling pathway that controls your temperature set point in the hypothalamus (Figure 6.23). Fever occurs when the set point is raised, making your body think it is too cold. Your body then sets into motion activities that raise temperature (chills, shivering, etc.).

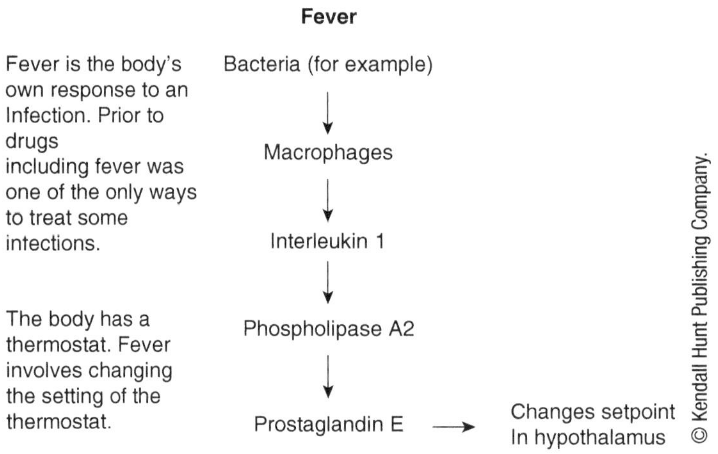

Figure 6.23. Fever in the body is induced by the presence of bacterial molecules that act on macrophages. Macrophages release a molecule called interleukin 1.

D. NSAIDs, Thromboxanes, and Platelets

Platelets are little packets of cytoplasm shed from megakaryocytes. They have no nucleus.

Thromboxanes play a critical role in platelet function and blood clotting. Platelets have receptors for a number of proteins and for thromboxane A2 that can trigger platelet activation. One of these proteins is collagen. Platelets are normally in the blood where blood vessels are lined with endothelial cells. Under these conditions, platelets normally will not activate. However, when there is damage to a blood vessel, platelets can be exposed to collagen. When a platelet encounter collagen, it activates.

Binding of collagen to the receptor on platelets triggers signaling events inside the platelet that do a number of things. One is to change the binding affinity of proteins on their surface known as integrins to bind to fibrin. A second event is the activation of phospholipase A2 which liberates arachidonic acid. The arachidonic acid is converted to thromboxane A2 which diffuses out of the platelet and activates neighboring platelets. The result is that many platelets clump together forming what is known as a platelet plug that prevents further bleeding. Platelets have often been likened to miniature smooth muscle and once they have formed a plug they proceed to contract, The contraction further closes the wound. Finally platelets secrete a growth factor into the surrounding area known as platelet-derived growth factor. This growth factor triggers certain types of cells known as fibrocytes to crawl into the area and make new connective tissue. The result of their activity is the formation of a scar.

We mentioned that NSAIDS block cyclooxygenase and therefor block the production of prostaglandins and thromboxanes. By blocking the formation of thromboxanes, NSAIDS block blood from clotting. Given that there are non-steroidal anti-inflammatory drugs we could probably guess that there are also steroidal anti-inflammatory drugs. Steroidal anti-inflammatory drugs act in the nucleus to turn on transcription of particular genes. An example of a steroidal antiinflammatory compound is cortisone, which can turn on the transcription of the lipcortin-1 gene. Lipocortin-1 inhibits phospholipase A2. Would you also expect it to block platelet clotting?

E. Blood, Plasma, and Serum

Whole blood contains blood cells, platelets, and serum proteins. When blood cells are removed from blood, the remaining platelet-containing fluid is called plasma. When whole blood is allowed to clot, the platelets react and the blood congeals into a solid mass. Then there is a phenomenon known as clot retraction, where the clot shrinks. Clot retraction is due to platelets contracting much the same as they do at the site of a wound. What is squeezed out is an amber-colored fluid that is called serum. Serum contains factors secreted by platelets that whole blood does not have.

During the early efforts at tissue culture, people tried to grow cells in a wide variety of nutrient media, even including whole blood in the mixture. Nothing worked. Eventually, serum was tried and it worked. Cells could be successfully grown in culture. Now serum is almost always used as a component of tissue culture media. Why do you suppose that serum worked but blood did not? It goes back to the platelets which secrete growth factors. As we will see, cells need nutrients to grow but they also need certain types of signals to trigger growth. We will delve into this more later in the course.

Reference

Shen, J. M., R. D. Li, and F. Y. Gao. 2005. "Effects of Ambient Temperature on Lipid and Fatty Acid Composition in the Oviparous Lizards, *Phrynocephalus przewalskii*." *Comparative Biochemistry and Physiology – Part B: Biochemistry & Molecular Biology* 142 (3): 293–301.

www.ingramcontent.com/pod-product-compliance
Ingram Content Group UK Ltd.
Pitfield, Milton Keynes, MK11 3LW, UK
UKHW050640211026
15997UKWH00024B/149

9 781524 981631